T0331002

Challenges and Sustainable Solutions in Bioremediation

This book highlights the characteristics, aims, and applications of microorganisms as a crucial solution for sustainable management of the toxic pollutants in aquatic ecosystems, soil, and air. It facilitates biotechnology towards the development of more sustainable biological systems by minimizing the level of harmful toxic substances and reducing the toxic effects in current chemical processes. It serves as a useful guide for a diverse community of practicing professionals, researchers, students (undergraduate and postgraduate), innovators involved in biotechnology research, and policymakers engaged in the development of strategies to deal with challenges of current environmental issues and working in the bioremediation field.

Features:

- Highlights the characteristics and applications of microorganisms as a crucial solution for sustainable management of pollutants in soil, water, and air.
- Examines how biotechnology can be used to remediate emerging toxic pollutants/contaminants from industrial wastewater and for nano-filtration applications.
- Discusses how bionanomaterials-based sensors can be practically used for monitoring air and water pollution, as well as resource recovery from wastewater.

Challenges and Sustainable Solutions in Bioremediation

Edited by
Kashyap Kumar Dubey, Ankush Yadav, and
Maulin P. Shah

CRC Press
Taylor & Francis Group
Boca Raton London New York

CRC Press is an imprint of the
Taylor & Francis Group, an **informa** business

Designed cover image: Shutterstock

First edition published 2025
by CRC Press
2385 NW Executive Center Drive, Suite 320, Boca Raton FL 33431

and by CRC Press
4 Park Square, Milton Park, Abingdon, Oxon, OX14 4RN

CRC Press is an imprint of Taylor & Francis Group, LLC

ISBN: 978-1-032-52581-5 (hbk)
ISBN: 978-1-032-52584-6 (pbk)
ISBN: 978-1-003-40731-7 (ebk)

DOI: 10.1201/9781003407317

Typeset in Times
by codeMantra

Contents

Preface

The continuously increasing level of industrialization and globalization has brought a lot of goods for the mankind; however, it also has resulted in several kinds of problems, in particular impacting the environment very badly. One of the major concerns in this regard has been the generation of hazardous contaminants as wastes from different sources (products of chemical nature and biomedical industries and manufacturers). The release of these wastes into the environment poses a serious impact on environmental health, and it has been a serious challenge to overcome this impact. Several methods have been developed in the past couple of decades for the removal/treatment of different emerging contaminants, in particular from wastewater and solid wastes. Among these, biological treatment processes have gained specific attention. The use of genetically modified microorganisms, in particular, has been of great significance for the treatment of the toxic and hazardous chemicals, which mineralize them to non-toxic forms.

The book *Challenges and Sustainable Solutions in Bioremediation* highlights the characteristics, aims, and applications of microorganisms and plants as a crucial solution for sustainable management of the toxic pollutants enriched in aquatic ecosystems, soil, and air. It facilitates treatment towards the development of more sustainable biological systems by minimizing the level of hazardous chemicals and reducing the toxic effects in current treatment processes. It also provides a comprehensive understanding of the subject knowledge with contemporary environmental challenges using bioremediation-based interventions. Readers will learn all about the recent, sustainable progress in theoretical, practical aspects and future potential applications of bio-systems in the environment.

In this book, a multidisciplinary team of researchers contributed to the challenges and sustainable solutions of bioremediation of emerging pollutants present in the different environmental matrices with their advancement and highlighted the technical, scientific, regulatory, safety, and societal impacts. The chapters focus on the different roles of biological techniques as remediation of emerging contaminants from the aquatic environment, eco-design and modification study of bio-nanoparticles, life-cycle assessment, nano-filtration, bionanomaterials-based sensors for monitoring air and water pollution, resource recovery from wastewater, and limitations for healthy and economically sustainable environment. Integration of biochemical principles, procedures, concepts, and attitudes into the current research and curriculum can form a robust platform and a powerful pedagogical foundation.

This book comprises ten chapters. Chapter 1 describes recent technologies and strategies for bioremediation of toxic and hazardous waste; Chapter 2 describes microbial biofilms: revolutionizing fermentation and bioremediation of environmental detoxification; Chapter 3 describes microbial strategies of chrysene remediation; Chapter 4 describes environmental remediation of noxious agrochemicals using engineered materials of bio-origin; Chapter 5 describes biochemical and molecular aspects of phytoremediation towards mitigation of heavy metals; Chapter 6 describes computational modelling in bioremediation: innovations and future directions;

Chapter 7 describes navigating nanomaterials: a path to sustainable wastewater treatment; Chapter 8 describes challenges and sustainable solution for the detection and bioremediation of microplastic pollution: challenges and scope of their remediation from the environment; Chapter 9 describes challenges in bioremediation: overcoming environmental and technological barriers; and Chapter 10 describes bioremediation strategies for petroleum hydrocarbons, heavy metals, and pesticides.

We hope that this book will be of immense usefulness to serve as reference literature for a diverse community of professionals, scientists, ecologists, environmental microbiologists, industrialists, researchers, students (undergraduate and postgraduate), innovators involved in bioremediation research, and policymakers who are engaged in the development of strategies to deal with challenges of current environmental issues and working in the bioremediation field.

We are grateful to the reviewers for their time and efforts in critically reviewing the manuscripts of the chapters. We sincerely acknowledge the CRC Press (Taylor & Francis Group) for giving us the opportunity to prepare this book under the agreement. We thank the team of Taylor & Francis Group: Ms. Maggie Apostolis, Editorial Assistant, CRS Press, and the entire team related with the processing of this book for their consistent support during the publication process.

Kashyap Kumar Dubey
Ankush Yadav
Maulin P. Shah

Editors' Profile

Prof. Kashyap Kumar Dubey is currently working as DEAN at the School of Biotechnology, JNU, New Delhi. His research interests are in biochemical engineering and wastewater treatment that includes process development of value-added pharmaceutical products through optimization of enzyme reactions and toxicological studies of micro-pollutants.

Dr. Ankush Yadav is currently working as Assistant Professor in the Department of Botany, Zakir Husain Delhi College, University of Delhi, New Delhi. His research interests are in wastewater treatment that includes the identification of hazardous emerging contaminants from wastewater; their toxicity analysis; and process development for bioremediation of emerging hazardous pharmaceuticals or biomedical waste from the aquatic environment.

Dr. Maulin P. Shah is an active researcher and scientific writer in his field for over 20 years. His research interests include biological wastewater treatment, environmental microbiology, biodegradation, bioremediation, and phytoremediation of environmental pollutants from industrial wastewaters. He has published more than 240 research papers in national and international journals of repute on various aspects of microbial biodegradation and bioremediation of environmental pollutants. He is the editor of 65 books of international repute.

List of Contributors

Anamika
Department of Chemistry
Awadhesh Pratap Singh University
Rewa, Madhya Pradesh, India

Anshu
Department of Zoology
CCS Haryana Agricultural University
Hisar, Haryana, India

Atul Arya
Department of Botany, Zakir Husain
    Delhi College
University of Delhi
New Delhi, India

Saroj Bala
Bioenergy Laboratory, Department
    of Renewable and Bio-Energy
    Engineering, College of Agricultural
    Engineering and Technology
Chaudhary Charan Singh Haryana
    Agricultural University
Hisar, Haryana, India

Srinivasan Balachandran
Bioenergy Laboratory, Department of
    Environmental Studies
Visva-Bharati
Santiniketan, West Bengal, India

Sandipan Banerjee
Mycology and Plant Pathology
    Laboratory, Department of Botany
Visva-Bharati
Santiniketan, West Bengal, India

Sneha Banerjee
Bioenergy Laboratory, Department of
    Environmental Studies
Visva-Bharati
Santiniketan, West Bengal, India

Sarita Dhaka
Department of Chemistry, Sanatan
    Dharma (PG) College
Maa Shakumbhari University
Saharanpur, Uttar Pradesh, India

Rahul Kumar Dhaka
Centre for Bio-Nanotechnology, College
    of Basic Sciences & Humanities,
    Chaudhary Charan Singh Haryana
    Agricultural University, Hisar, India
and
Department of Chemistry, Centre for
    Bio-Nanotechnology, College of
    Basic Sciences & Humanities
Chaudhary Charan Singh Haryana
    Agricultural University
Hisar, Haryana, India

Sushmita Gandash
Department of Chemistry
Suraj Degree College
Gurugram, Haryana, India

Rajeeva Gaur
Department of Microbiology
Dr. Rammanohar Lohia Avadh
    University
Ayodhya, Uttar Pradesh, India

Anudeb Ghosh
Bioenergy Laboratory, Department of
   Environmental Studies
Visva-Bharati
Santiniketan, West Bengal, India

Nitu Gupta
Department of Environmental Science
Tezpur University
Napaam, Tezpur, Assam, India

Raza Rafiqul Hoque
Department of Environmental Science
Tezpur University
Napaam, Tezpur, Assam, India

Dolly Kain
Department of Botany, Deen Dayal
   Upadhyaya College
University of Delhi
New Delhi, India

Apurba Koley
Bioenergy Laboratory, Department of
   Environmental Studies
Visva-Bharati
Santiniketan, West Bengal, India

Akshay Kumar
School of Biotechnology
Jawaharlal Nehru University
New Delhi, India

Ashwani Kumar
Department of Nutrition Biology
Central University of Haryana
Mahendergarh, Haryana, India

Pradeep Kumar
Department of Vocational Studies &
   Skill Development
Central University of Haryana
Mahendergarh, Haryana, India

Rahul Kumar
Department of Zoology, CCS Haryana
   Agricultural University
College of Agriculture
Bawal, Rewari, Haryana, India

Ravi Kumar
Department of Processing and Food
   Engineering, College of Agricultural
   Engineering and Technology
Chaudhary Charan Singh Haryana
   Agricultural University
Hisar, Haryana, India

Satender Kumar
Department of Soil Science, College of
   Agriculture
CCS Haryana Agricultural University
Hisar, Haryana, India

Suresh Kumar
Department of Botany, Ramjas College
University of Delhi
New Delhi, India

Naseeb
Department of Vocational Studies &
   Skill Development
Central University of Haryana
Mahendergarh, Haryana, India

Nisha
Department of Biochemistry
Kurukshetra University
Kurukshetra, Haryana, India

Manvendra Patel
School of Environmental Sciences
Jawaharlal Nehru University
New Delhi, India

Anuj Rana
Department of Microbiology, Centre
    for Bio-Nanotechnology, College of
    Basic Sciences & Humanities
Chaudhary Charan Singh Haryana
    Agricultural University
Hisar, Haryana, India
and
Centre for Bio-Nanotechnology, College
    of Basic Sciences & Humanities
Chaudhary Charan Singh Haryana
    Agricultural University
Hisar, Haryana, India

Swati Rani
Department of Bio & Nanotechnology
Guru Jambheshwar University of
    Science and Technology
Hisar, Haryana, India

Rajni Sharma
Department of Biotechnology
Maharaja Agrasen University
Baddi, Himachal Pradesh, India

Binoy Kumar Show
Bioenergy Laboratory, Department of
    Environmental Studies
Visva-Bharati
Santiniketan, West Bengal, India

Adarsh Kumar Shukla
Department of Genomics Research
Sri Sathya Sai Sanjeevani Research
    Foundation
Palwal, Haryana, India
Department of Nutrition Biology
Central University of Haryana
Mahendergarh, Haryana, India

Gulab Singh
Department of Bio-Nanotechnology,
    College of Biotechnology
CCS, Haryana Agriculture University
Hisar, Haryana, India

Harsha Singh
Medicinal Plant Research Laboratory,
    Department of Botany, Ramjas
    College
University of Delhi
New Delhi, India

Love Singla
Department of Microbiology
Maharaja Agrasen University
Baddi, Himachal Pradesh, India

Manikant Tripathi
Department of Biotechnology
Dr. Rammanohar Lohia Avadh
    University
Ayodhya, Uttar Pradesh, India

Ankush Yadav
Department of Botany, Zakir Husain
    Delhi College
University of Delhi
New Delhi, India

1 Recent Technologies and Strategies for Bioremediation of Toxic and Hazardous Waste

Nisha, Anshu, Swati Rani, Satender Kumar, Rahul Kumar, and Akshay Kumar

1.1 INTRODUCTION

Global industrialization, rising anthropogenic activities, unregulated and unsafe agricultural practices, and rising population have all contributed to an increase in environmental pollution and topsoil pollution over the past few decades. These factors have led to the accumulation of hazardous waste, including plastic, metal, and other materials trash from industry, agriculture, and residential use, including rubber, metal, and plastic (Kumar et al., 2020; 2022a; 2022b). The water quality has declined due to human activity and the ultimate dispose of hazardous metals, chemicals, and elements from the steel, energy, and battery industries and this had raised major environmental concerns.

By disposing of toxic wastes and producing a significant amount of heavy metals, chemicals contaminate the soil (Yadav et al., 2022a; 2022b), and various industries like carpet, textile, and petrochemical production seriously harm the environment. This presents a significant challenge for harmful treatment practices (Farhadian et al., 2008). Contaminated soil from agricultural or industrial processes offers substantial health risks to people and animals, and it can harm the environment and living organisms in particular areas and finally cause economic loss. Hydrocarbon concentrations (40%–60%), water (30%–90%), and mineral particle concentrations (5%–40%) vary in waste from the petroleum sector. Hazardous and non-hazardous garbage are separated from them (Kumar et al., 2023; 2024).

Hazardous waste is defined as any waste that is corrosive, explosive, combustible, reactive, inflammable, or poisonous. The main cause of environmental degradation that needs to be eradicated is the continued dumping of synthetic harmful chemicals like herbicides, insecticides, plastics, fertilizers, and things that include hydrocarbons. Concerns about the environment and health have significantly increased. Heavy metals are contaminants that naturally occur in the crust of the Earth and are difficult to break down. Heavy metals in soil decrease the quality and amount of food by preventing plant development, nutrient uptake, and physiological metabolic

DOI: 10.1201/9781003407317-1

processes (Masindi et al., 2021; Yadav et al., 2023). Chemical, biological, and physical techniques are used to clean up soils that have been contaminated with metal (Rebello et al., 2021). Environmental toxins can be eliminated using a realistic and affordable method called bioremediation (Tripathi et al., 2021). An efficient, secure, and less expensive technique for removing contaminants is to use microorganisms and their enzymes (Karigar and Rao, 2011; Kumar and Singh 2022). According to Kadri et al. (2017), in the bioremediation process, chemical reactions occur by a variety of enzyme types, including oxidoreductases, hydrolases, laccases, and peroxidases. In a study, endophytic fungi and *Aspergillus sydowii* were utilized to remove several contaminants using bioremediation technology, including organophosphate insecticides (Soares et al., 2021). In order to treat contaminated sites and restore them to their pre-contamination state, bioremediation uses primarily microorganisms, living organisms, plants, and their enzymes to eliminate, mineralize, transform, and degrade the perilous elements (Azubuike et al., 2016). Co-metabolism, biotransformation kinetics, biotreatment, and biogeochemical modeling are core study areas in the subject of bioremediation and their main research areas include biochemical evaluation methods, co-metabolic methods, fate modeling, and ecological sustainability (Harekrushna and Kumar, 2012; Kumar et al., 2022b). Several technologies, including nanotechnology, biochemical and genetic engineering, chromatography, as well as spectrometry, are used to carry out and monitor the bioremediation processes (Figure 1.1). Science is now paying attention to bioluminescence-based assays for the eco-toxicological evaluation of contaminants. Microbiological biosensors have also

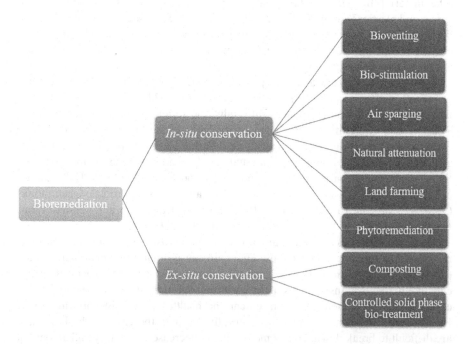

FIGURE 1.1 Types of bioremediation.

been investigated as a friendly and trustworthy method for locating and monitoring the pollutants within an environment (Bose et al., 2021). Nanotechnology is gaining interest in the realm of environmental management and protecting the environment from dangerous substances. Bioremediation is widely regarded as an "environmentally appropriate" method because it is so clean, green, and sustainable. Since the recent development of biomimetic nanotechnology, researchers have been drawn to thoroughly investigate this unexplored path.

Nanotechnology can lower the amount of money that businesses spend on environmentally beneficial ways to decrease these pollutants. In contrast to their bulk phase, nanoscale products have a bigger surface area and higher efficiency of reactivity, thanks to the burgeoning field of nanotechnology. Nanoparticles' distinctive characteristics offer various potential beneficial attributes to decontaminate places, contaminated with metals, herbicides, and petroleum hydrocarbons. The bioremediation of polluted environments has been receiving more and more attention due to its self-propelling, economical, and environmentally favorable qualities (Benjamin et al., 2019). For example, *Escherichia coli* have the capacity to make copper-resistant nanoparticles, and these nanoparticles are capable of breaking down azo dye and textile effluents. Additionally, the use of these nanoparticles to treat industrial effluent results in treated samples having less suspended particulates (Noman et al., 2020). Quantum confinement, or the confinement of electrons within particles with dimensions smaller than the bulk electron delocalization length, is responsible for the peculiar physicochemical and optoelectronic features of nanoparticles. In this approach, bioremediation and nanobiotechnology can be combined to create a cost-effective, long-lasting solution for a clean, green environment.

1.2 PRINCIPLE OF BIOREMEDIATION

"Bioremediation" is the word used to describe the biological degradation of organic wastes under regulated conditions (Figure 1.2). Through the use of bioremediation, toxic compounds can be broken down or detoxified by giving organisms the nutrients and other chemicals they require to function at their best (Chen et al., 2016). For successful bioremediation, enzymatic reaction of microorganisms must occur at favorable condition for sustainable development (Kumar et al., 2020). When this microbial growth continues to occur, a harmless surrounding environment may develop within the time (Kumar and Singh 2022). However, its application frequently entails changing environmental parameters to promote microbial growth and degradation. Every stage of the metabolic process requires the use of enzymes (Chen et al., 2016; Malik et al., 2022). It belongs to the same family as hydrolases, lyases, transferases, and oxidoreductases. For bioremediation to be effective, the contaminants must be subjected to enzymatic activity. Bioremediation techniques can remediate some pollutants on-site, minimizing exposure hazards for clean-up workers or potentially wider exposure as a result of transportation accidents. They are often more cost-effective than conventional procedures like cremation. The most important step for bioremediation technology is biodegradation. It involves transforming dangerous organic pollutants, like as carbon dioxide and water, into harmless, naturally occurring inorganic substances that can be used by people, plants, animals, and aquatic life.

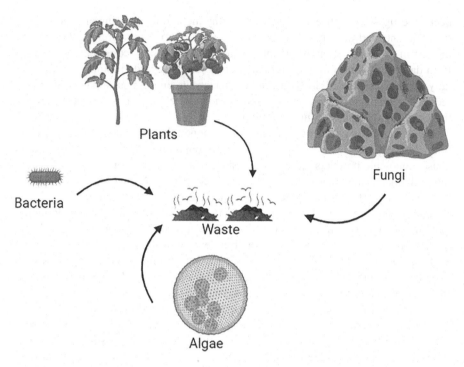

FIGURE 1.2 Diagrammatic representation of the principle of bioremediation.

The public views bioremediation as being more acceptable than other solutions because it is based on natural attenuation (Priyadarshanee and Das, 2021).

1.3 RECENT ADVANCEMENT IN BIOREMEDIATION

Nanotechnology is a multidisciplinary field of science and engineering that deals with materials and devices at the nanoscale, typically involving structures and phenomena with dimensions on the order of nanometers. Although it is a relatively new field of study, nanotechnology has been practically employed by humans for thousands of years in its contemporary and postmodern forms (Anuradha, 2013). Nanotechnology encompasses manipulating, controlling, and utilizing materials and structures at the nanoscale to create the latest properties, functions, and applications. It involves working with individual molecules, atoms, or nanoparticles to engineer materials with unique and tailored properties. The nanoscale refers to a range between 1 and 100 nm, where unique and often extraordinary properties and behaviors emerge due to the minimal size. Based on their catalytic, magnetic, electrical, and optical capabilities, the synthesized nanoparticles with higher surface energy, greater surface fraction, ideal dimensions, and spatial confinements were determined to have prospective benefits in the field. The field of nanotechnology does include bio-based nanoparticles, which have the potential to help environmental protection through, among other things, contaminant removal, treatment, and pollution prevention

(Bala et al., 2022). Consequently, one of the rapidly expanding study interests in technological advancement in recent years is environmental nanobiotechnology.

1.3.1 Nanotechnology for Remediation of Heavy Metal

The different nanoparticles with desirable properties and their application in wastewater treatment for the removal of heavy metals and other contaminants are discussed in many research articles. Among all the water contaminants, heavy metals are a major contributor to the serious health issues that they have on both people and animals. Heavy metal pollution of soil and groundwater has become a serious environmental issue on a global scale. Because these heavy metals can build up in the food chain and have an impact on the health of living things, it is crucial that they are removed from polluted soil and groundwater (Tyagi and Kumar, 2021). This is because of their extreme toxicity and lack of biodegradability. A well-known illness in Japan called Itai-Itai is brought on by prolonged exposure to the toxic metal chromium. Chromium causes severe and protracted illnesses like hypertension, fetal skeletal malformations, testicular atrophy, renal damage, and emphysema. In order to completely remove these dangerous metals from water, new, creative approaches are thus imperative. To adsorb heavy metal ions from aqueous solutions, various types of novel materials are being researched, including graphene derivatives, carbon-based sorbents, chelates, activated carbons, chitosan/natural zeolites, and clay minerals. According to report, the plant species *Brachiaria mutica* and *Zea mays* were used in a phytoremediation technique to treat a heavy metal-contaminated site.

1.3.2 Applications of Microorganisms-Assisted Nanotechnology

Although many traditional physicochemical techniques are currently used to remediate environmental contaminants, including precipitation, electrochemical treatment, electrocoagulation, and adsorption, there is still a pressing need for the development of novel, efficient, environmentally friendly, and economically viable methods (Fomina and Gadd, 2014). Biofabrication, also known as the incorporation of microorganisms into nanotechnology processes, is a sustainable and environmentally friendly strategy. The use of chemicals and self-agglomeration in aqueous solutions are potential drawbacks of conventional chemical methods for producing nanoparticles. As a result, a promising solution has been found in the environmentally friendly synthesis of nanoparticles using plant extracts, fungi, and bacterial enzymes. These bacteria act like reducing agents for metal complex salts, which help to produce metallic nanoparticles. This enhanced stability in aqueous environments is a result of co-precipitation or the incorporation of proteinaceous and bioactive components onto the surfaces of the biogenically created nanoparticles. For instance, *Aspergillus tubingensis* (STSP 25), which was isolated from the rhizosphere of *Avicennia officinalis* in India, was used to biofabricate iron oxide nanoparticles in the study by Mahanty et al. (2020). These nanoparticles (Pb^{2+}, Ni^{2+}, Cu^{2+}, and Zn^{2+}) showed the ability to efficiently remove over 90% of heavy metals from wastewater with the added benefit of renewability for up to several cycles. Through endothermic reactions, the metal ions are chemically adsorbed on surfaces of the nanoparticles.

Exopolysaccharides (EPS) from *Chlorella vulgaris* were used in a different study to co-precipitate iron oxide nanoparticles. An analysis using Fourier-transform infrared spectroscopy (FT-IR) proved that adding functional groups from EPS to nanoparticles was successful. Additionally, as documented by Govarthanan et al. (2020), the resulting nanocomposite demonstrated remarkable removal abilities, removing 91% of PO_4^{3-} and 85% of NH_4^+ from wastewater. The synthesis of nanoparticles using microorganisms has made environmentally friendly, low-cost methods possible. Copper nanoparticles were produced using a strain of *Escherichia* species, SINT7, which is known for its ability to withstand copper. According to the study by Noman et al. (2020), these biogenic nanoparticles showed the ability to degrade azo dyes and textile effluents, resulting in significant reductions in various dye concentrations and industrial effluent parameters.

Additionally, *Pseudoalteromonas* species CF10–13's use in the preparation of the iron–sulfur nanoparticles provided an environmentally friendly method of biodegradation by minimizing the production of harmful gases and metal complexes. These nanoparticles efficiently degraded the Naphthol Green B dye through extracellular electron transfer. The treatment of industrial effluents with this biogenic nanoparticle technology has great potential to be both affordable and long-lasting. Microorganisms provide a number of additional ways to advance nanotechnology in addition to directly producing nanoparticles. One such method involves using microorganisms that produce catalytic enzymes in combination with nanoparticles to effectively remediate effluents. Due to their diverse functions, microorganisms offer a superior technology for addressing environmental issues in a variety of industries.

1.3.3 GENETICALLY ENGINEERED MICROORGANISM-BASED BIOREMEDIATION

Nanotechnology applications within the realm of environmental biotechnology offer promising innovations, with potential cross-pollination from more rapidly evolving fields like medical nanotechnology. A noteworthy avenue of exploration involves the functionalization of nanostructures with biomolecules. For instance, in a study by Bolisetty and Mezzenga (2016), membranes incorporating amyloid proteins and activated porous carbon were devised to remove and recover heavy metals. Inspired by amyloid protein formation in neurons, the researchers transformed milk proteins into amyloid fibrils, which effectively captured various ions. This research underscored the significance of identifying cost-effective sources of biomolecules for such advancements. Natural proteins can be produced inexpensively through developed recombinant technologies and feature 20 different amino acids, enabling diverse interactions with other molecules and the creation of novel catalytic surfaces and structures (Ljubetič et al., 2017). Biotechnology also plays a role in providing eco-friendly methods for functionalizing nanoparticles. For example, Gao et al. (2019) introduced the biological toolbox by incorporating the bacterium *Komagataeibacter sucrofermentans* to create specially modified cellulose-like polymers. In order to grow these bacteria, conventional bioreactor techniques are used and glucose monomers containing the desired chemical modifications are fed to the bacteria. These modifications are then biologically incorporated into the polymer. This method avoids the use of difficult solvents, stoichiometry, and the creation of residues that are harmful to the

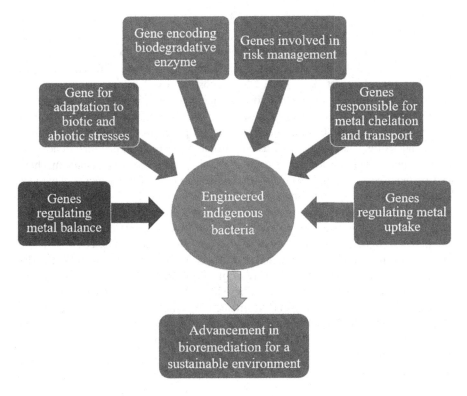

FIGURE 1.3 Genetically modified bacteria for improved bioremediation, a greener environment, and sustainability.

environment (Figure 1.3). The synthesis of cellulose-based nanomaterials has a significant potential for simplification through improvements in traditional and cutting-edge biotechnological techniques (such as mutagenesis, protein engineering, or gene editing) for a variety of applications by Sharma et al. (2019).

The recent advancement of RNA-based fungicides raises the prospect of an effective replacement for conventional biochemical fungicides. In order to silence the genes of fungal pathogens, double-strand RNAs are applied to leaves or fruits in the expectation that they can hybridize with their essential mRNAs given by Wang et al. (2016). Biologically derived capsules had the best soil penetration and cargo release, according to Chariou et al. (2019), who tested conventional silica particles, recombinant or plant-produced virus capsules, and other nanoscale encapsulation architectures. These biodegradable bionanoparticles are eco-friendly because they don't produce any organic waste. The next generation of nanoparticle biofunctionalization is anticipated to have a significant positive impact on this field.

One underexplored frontier in environmental biotechnology is the three-dimensional construction of DNA structures via DNA hybridization. The relatively straightforward yet adaptable rules governing nucleotide recognition between different DNA strands can be harnessed to build diverse geometric configurations, referred to as "molecular origami," starting with basic crossover tiles and ending with polyhedral meshes. As demonstrated by the development of DNA nanorobots that can transport challenging molecules that are like other DNAs/RNAs or proteins and use them as cargo, fasteners, or springs, these molecular structures offer an innovative array of functionalization tools. In a study by Li et al. (2018), a DNA nanosheet was developed to interact with the protein known as thrombin and compartmentalize it until other DNA molecules could seal the nanosheet and act as fasteners. The DNA origami, transformed into a nanotube, would only open in the presence of a particular key, such as the tumor protein nucleolin, leading to the release of thrombin and the coagulation and necrosis of the tumor. This innovation demonstrates the programmable and complex mechanics that biomolecular-based biorobots are efficient in case of biomedicine. This idea has consequences for environmental applications such as the development of new pesticides or the elimination of pathogenic bacteria which are tolerant to antibiotics.

It is an emerging discipline to use oxygen-sensitive proteins identified in people and plants to make oxygen biological sensors and inducible genetic circuits. These can be modified to create functional nanomaterials capable of stoichiometrically responding to oxygen levels. They were originally designed for in vivo applications given by Vázquez-Núñez et al. (2020). These ideas, which have been tested with proteins, are also shown to work with DNA molecules, which can filter out complex analytes like proteins and provide a variety of sensing and purification options. Recent reviews by Ryu et al. (2019) have focused on the incorporation of transmembrane proteins into membranes and their coupling to different transducers for applications such as pesticide detection, gas monitoring, energy harvesting, and microarrays.

1.3.4   BIOREMEDIATION USING BIOSENSORS

1.3.4.1   Nanomaterial-Derived from Electrochemical Biosensors

Different pronounced advantages are offered by electrochemical sensors. These gadgets provide sophisticated ways to link molecular signal transduction processes and biological recognition events, in particular. Additionally, due to their exceptional suitability for decentralized testing in terms of size, cost, low volume, and power, electrochemical devices hold great promise for a variety of applications in the biomedical and environmental fields. The effective implementation of these sensing protocols depends on the controlled surface architectures created by the structured arrangement of nanomaterials.

1.3.4.2   Enzyme Electrodes Utilizing Nanomaterials

Numerous clinically and environmentally important substrates can be monitored using enzyme electrodes. Establishing reliable electrical communication between the enzyme's active site and the electrode surface is a crucial challenge in amperometric enzyme electrodes. Examples include the use of self-assembled "forests" of aligned carbon nanotubes (CNTs) as molecular wires to enable electrical communication between the underlying electrode and redox proteins attached to the Single Walled Carbon Nanotube (SWCNT) ends. The "plugging" of an electrode into glucose oxidase is made possible by this method. Furthermore, metal nanoparticles, like gold nanoparticles, have shown value as electron relays for positioning glucose oxidase on conducting supports and connecting its redox center, enhancing electron transport. Carbon nanotube–modified electrodes encourage electron transfer reactions and reduce surface fouling problems related to the oxidation of liberated Nicotinamide Adenine Dinucleotide Hydrogen (NADH) because of edge plane defects at their end caps.

The detection of peroxide species released by enzymes has also been improved by the deposit of platinum nanoparticles onto CNTs. In addition to CNT films, CNT-based inks and pastes have been used to create biocomposite and screen-printed amperometric biosensors. Metal nanoparticles have better electrocatalytic qualities than bulk metal electrodes, which is advantageous for amperometric enzyme electrodes. For instance, the amperometric biosensing of glutamate has been improved by iridium nanoparticles dispersed in carbon that resembles graphite. A pair of nanoelectrodes separated by a polyaniline/glucose oxidase film have also been used in conducting polymer nanosensors for glucose detection, which have a quick response time and little oxygen consumption.

1.4 BIOSURFACTANTS

Biosurfactants are naturally occurring molecules produced by microorganisms, plants, or animals. They have surfactant properties, which means they can reduce surface tension. Biosurfactants have different attributes across various industries due to their eco-friendly and biodegradable nature. Some common applications of biosurfactants include:

1. **Enhanced Oil Recovery (EOR)**: Oil and gas companies use biosurfactants to increase oil recovery from reservoirs. They ease the displacement of oil trapped in rock formations by lowering the interfacial tension between water and oil.
2. **Bioremediation**: Biosurfactants aid in the bioremediation of contaminated sites. They can increase the bioavailability of hydrophobic contaminants like petroleum hydrocarbons, facilitating their degradation by microorganisms.
3. **Detergents and Cleaning Agents**: Biosurfactants are used in the formulation of eco-friendly and biodegradable detergents and cleaning agents. They help remove grease and oil from surfaces effectively.
4. **Agriculture**: Biosurfactants can be used as adjuvants in pesticide formulations. They improve the wetting and spreading of pesticides on plant surfaces, increasing their efficacy.
5. **Food Industry**: Biosurfactants are employed in food processing and as food additives to enhance the texture and stability of food products. They can be used in emulsification, foaming, and as antimicrobial agents.
6. **Cosmetics**: They are used in the formulation of cosmetics and personal care products, such as shampoos, creams, and lotions, to improve the texture and stability of these products.
7. **Phytoremediation**: Biosurfactants can enhance the uptake of contaminants by plants in phytoremediation processes, improving the removal of pollutants from soil and water.

8. **Biomedical Applications**: Biosurfactants have potential applications in medical and pharmaceutical industries, such as wound care and drug delivery systems.

9. **Enhanced Nutrient Uptake in Agriculture**: Biosurfactants can improve nutrient uptake by plants in agriculture, potentially leading to increased crop yields.

10. **Biomineralization**: They are used in processes like biomineralization, where they influence the formation of minerals and nanoparticles for various industrial applications.

11. **Environmental Remediation**: Biosurfactants can aid in the cleanup of oil spills and other environmental disasters by breaking down and dispersing oil and contaminants.

12. **Microbial Enhanced Oil Recovery (MEOR)**: In MEOR processes, biosurfactant-producing microorganisms are injected into oil reservoirs to enhance oil recovery.

1.4.1 APPLICATIONS OF NANO-ADSORBENTS AND NANOFILTRATION MEMBRANES

There is widespread acceptance of nanoparticles as successful adsorbents for removing dangerous contaminants from industrial wastewater. According to the research by Kumari et al. (2019), organic and inorganic pollutants are removed by the use of nano-adsorbents. This type of nanoparticles can be broadly divided into three groups: metal, metal oxide, and carbon-based nanoparticles. Notably, Kumari et al. (2019) describe carbon-based nanoparticles. For example, carbon nanotubes are excellent adsorbents for hazardous substances in industrial and pharmaceutical wastewater. According to Takmil et al. (2020), activated carbon-modified nanomagnets are further proven to be effective at removing fluoride ions from wastewater, with a remarkable removal rate of 97.4%. Additionally, nanocatalysts and microbial fuel cells have been used to generate bioelectricity. Notably, electrodes coated with iron(II) molybdate nanocomposites significantly increased microbial fuel cell efficiency, as shown by the work of Mohamed et al. (2020), which resulted in a maximum columbic efficiency of 21.3 0.5%, a power density of 106 3 mW/m^2, and a Chemical oxygen demand (COD) removal competence of 79.8 1.5%. Chromium can be removed from wastewater using superparamagnetic composites made of iron oxide nanoparticles and activated carbon that pass Environmental Protection Agency (EPA) standards for discharge, using the methods of magnetic separation and sorption, as stated by Nogueira et al. (2019). A graphene-based nanocollector was also introduced by Hoseinian et al. (2020). Through an affordable, effective, and stable ion flotation process, this amino-functionalized graphene oxide nanocollector removed nickel ions from wastewater almost entirely. Metal and metal oxide-based nano-adsorbents are essential for removing pollutants from wastewater, in addition to carbon-based nanoparticles. According to research, coating magnetic nanoparticles with additional substances improves the effectiveness of their adsorption. According to Najafpoor et al. (2020), magnetic nanoparticles coated with silver removed 36.56% more COD from wastewater than uncoated magnetic nanoparticles.

1.5 MICROBIAL BIOREMEDIATION

Microbial bioremediation is a valuable tool in environmental remediation efforts, offering a sustainable and often cost-effective way to clean contaminated sites and mitigate environmental damage. This process employs microorganisms, like bacteria, fungi, and algae, to degrade or eradicate pollutants from contaminated environments. The term "autochthonous microorganisms" refers to those that naturally live in ecosystems of soil or water that are experiencing a separation, or by other bacteria that are derived from different habitats can be used to carry out bioremediation (Verma and Kuila, 2019). Rehabilitation and detoxification using microorganisms of contaminated soil have shown to be the most efficient, simple, and secure technology. In the environment, hazardous substances have been generated as a consequence of human activities, including the production of fuel; industrial operations; ore mining; oil and gas extraction; and the use of organic solvents, pesticides, pigments, and plastics. Native soil bacteria can remove or detoxify these substances (Garbisu and Alkorta, 2001.) Here are some key points about microbial bioremediation.

1.6 BIOREMEDIATION BY BIOAUGMENTATION

Bioaugmentation is introducing specific microorganisms or biological materials into environments like agriculture, wastewater treatment, and environmental remediation to enhance natural processes, improving efficiency. The goal of bioaugmentation is to improve the efficiency or effectiveness of biological processes by introducing organisms or substances that can perform desired functions. Bioaugmentation is considered an eco-friendly and sustainable approach because it relies on natural processes and reduces the need for harsh chemicals or mechanical interventions (Figure 1.4). This process frequently involves the introduction of microbes, any native or exogenous, to the contaminated locations. It is a remediation procedure that is economical, efficient, and quick, and it is becoming more and more popular with site managers and remediation experts (Ivask et al., 2011).

1.7 ADVANTAGES OF BIOREMEDIATION

1. **Environmentally Friendly:** Bioremediation relies on natural processes and microorganisms to degrade contaminants, making it environmentally sustainable and less disruptive than many chemical or mechanical methods.
2. **Minimizes Soil Disruption:** In situ bioremediation methods, such as bioventing and phytoremediation, allow for treatment without the need to excavate and transport contaminated soil, minimizing disruption to the site.
3. **Selective for Target Contaminants:** Microorganisms can be engineered or selected to target specific contaminants, making bioremediation highly effective for certain pollutants.
4. **Reduced Secondary Contamination Risk:** Unlike some chemical methods, bioremediation generally reduces the risk of generating secondary contaminants.

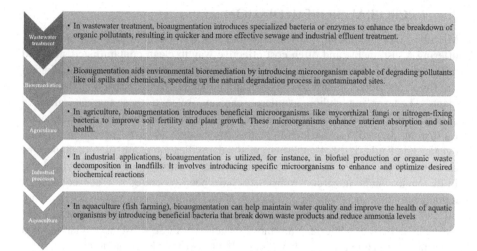

FIGURE 1.4   Key areas of applied bioaugmentation.

1.8   DISADVANTAGES OF BIOREMEDIATION

1. **Time-Consuming:** Bioremediation is a slow process, especially for complex or highly contaminated sites, which may take years or even decades for complete clearout.
2. **Site-Specific:** The effectiveness of bioremediation can vary depending on site-specific conditions like pH, temperature, and the existence of suitable microorganisms.
3. **Limited Applicability:** Bioremediation may not be suitable for all types of contaminants or for sites with extreme conditions, such as high levels of radioactivity or extremely acidic or alkaline soils.
4. **Uncertainty:** The outcome of bioremediation can be less predictable than some chemical treatments, and there can be uncertainty regarding the timeline and success of the process.
5. **Public Perception:** There may be public concerns about the usage of genetically modified organisms (GMOs) or the release of potentially harmful microorganisms into the environment.

1.9   FUTURE PERSPECTIVES

The genetics and metabolic pathways of microbes that break down pollutants are now well understood because of the developments in genomics, metagenomics, and synthetic biology. With the help of this understanding, natural biodegradation processes can be improved and changed, potentially increasing the efficacy and efficiency of bioremediation. In order to maximize bioremediation, scientists are looking into how to engineer the microbial communities found in contaminated sites. To improve the capacity for pollutant breakdown, particular microbes must be added or altered.

Technology for bioreactors is developing to offer regulated settings for bioremediation. This includes the creation of novel bioreactor designs that could be applied in situ or ex situ to the treatment of wastewater, groundwater, and contaminated soil. Bioremediation techniques are being made more effective by the use of nanomaterials. The bioavailability and breakdown rates of specific pollutants can be increased by using nanoparticles as transporters for nutrients and microbes. A contaminated location can be bioaugmented by adding particular microorganisms with the necessary pollutant-degrading capability. Biostimulation includes supplying nutrients or other growth-promoting substances to promote the activity of microorganisms. Both tactics are green chemistry: Efforts are made to develop eco-friendly and sustainable bioremediation approaches that minimize the environmental impact of the remediation process itself. Green chemistry principles are guiding the design of bioremediation strategies. Support for these eco-friendly alternatives can be gained by educating communities and the general public about bioremediation processes and their advantages.

1.10 CONCLUSIONS

A significant issue is the bioremediation of toxic chemicals from the ecosystem because heavy metals are primarily regarded as toxic substances and require immediate remediation. Several technologies have addressed the removal of toxic pollutants. Recent technologies and strategies in bioremediation offer a promising path forward in addressing toxic and hazardous waste contamination. The synergistic use of bioaugmentation, phytoremediation, and natural attenuation, among others, optimizes the withdrawal of hazardous substances from the ecosystem. A sustainable approach to the biological remediation of industrial effluents was made possible by integrating nanotechnology and microorganisms. Microorganism-infused nanomaterials make it feasible. An efficient and resilient remediation process may result from microbiome engineering, which modifies microbial populations' makeup to maximize their biodegradation capacity. Additionally, it has been reported that microorganisms can produce nanomaterials, opening up more effective and persistent options for effluent remediation. Developing technologies like nanotechnology and remote sensing have enhanced the monitoring and control of bioremediation processes. Similarly, nanoparticles can be used to improve the delivery of nutrients or electron donors to microbial populations, enhancing their biodegradation capabilities. Genetically engineered microorganisms that efficiently metabolize a broad range of toxic substances are developed. These engineered microbes have shown remarkable effectiveness in breaking down pollutants like hydrocarbons, heavy metals, and persistent organic compounds. Enzyme nanotechnology has developed long-lasting, highly active, stable enzymes with various applications like immobilization of several enzymes on a single nanomaterial carrier which make it possible to design customized enzyme systems that collaborate to break down intricate contaminant mixtures. These holistic approaches remove contaminants and support long-term ecosystem health and resilience. More commercial research should be done to utilize this method fully.

REFERENCES

Anuradha, J., 2013. *Use of Some Aquatic and Terrestrial Weeds in the 'Green' Synthesis of Gold Nanoparticles* (Doctoral dissertation), Pondicherry University, India.

Azubuike, C.C., Chikere, C.B. and Okpokwasili, G.C., 2016. Bioremediation techniques–classification based on site of application: principles, advantages, limitations and prospects. *World Journal of Microbiology and Biotechnology*, 32, pp. 1–18.

Bala, S., Garg, D., Thirumalesh, B.V., Sharma, M., Sridhar, K., Inbaraj, B.S. and Tripathi, M., 2022. Recent strategies for bioremediation of emerging pollutants: a review for a green and sustainable environment. *Toxics*, 10(8), p. 484.

Benjamin, S.R., Lima, F.D., Florean, E.O.P.T. and Guedes, M.I.F., 2019. Current trends in nanotechnology for bioremediation. *International Journal of Environment and Pollution*, 66(1–3), pp. 19–40.

Bolisetty, S. and Mezzenga, R., 2016. Amyloid–carbon hybrid membranes for universal water purification. *Nature Nanotechnology*, 11(4), pp. 365–371.

Bose, S., Kumar, P.S. and Vo, D.V.N., 2021. A review on the microbial degradation of chlorpyrifos and its metabolite TCP. *Chemosphere*, 283, p. 131447.

Chariou, P.L., Dogan, A.B., Welsh, A.G., Saidel, G.M., Baskaran, H. and Steinmetz, N.F., 2019. Soil mobility of synthetic and virus-based model nanopesticides. *Nature Nanotechnology*, 14(7), pp. 712–718.

Chen, B.Y., Ma, C.M., Han, K., Yueh, P.L., Qin, L.J. and Hsueh, C.C., 2016. Influence of textile dye and decolorized metabolites on microbial fuel cell-assisted bioremediation. *Bioresource Technology*, 200, pp. 1033–1038.

Farhadian, M., Vachelard, C., Duchez, D. and Larroche, C., 2008. In situ bioremediation of monoaromatic pollutants in groundwater: a review. *Bioresource Technology*, 99(13), pp. 5296–5308.

Fomina, M. and Gadd, G.M., 2014. Biosorption: current perspectives on concept, definition and application. *Bioresource Technology*, 160, pp. 3–14.

Gao, M., Li, J., Bao, Z., Hu, M., Nian, R., Feng, D., An, D., Li, X., Xian, M. and Zhang, H., 2019. A natural in situ fabrication method of functional bacterial cellulose using a microorganism. *Nature Communications*, 10(1), p. 437.

Garbisu, C. and Alkorta, I., 2001. Phytoextraction: a cost-effective plant-based technology for the removal of metals from the environment. *Bioresource Technology* 77, pp. 229–236.

Govarthanan, M., Jeon, C.H., Jeon, Y.H., Kwon, J.H., Bae, H. and Kim, W., 2020. Non-toxic nano approach for wastewater treatment using Chlorella vulgaris exopolysaccharides immobilized in iron-magnetic nanoparticles. *International Journal of Biological Macromolecules*, 162, pp. 1241–1249.

Harekrushna, S. and Kumar, D.C., 2012. A review on: bioremediation. *International Journal of Environmental Research*, 2(1), pp. 13–21.

Ivask, A., Dubourguier, H.C., Põllumaa, L. and Kahru, A., 2011. Bioavailability of Cd in 110 polluted topsoils to recombinant bioluminescent sensor bacteria: effect of soil particulate matter. *Journal of Soils and Sediments*, 11, pp. 231–237.

Kadri, T., Rouissi, T., Brar, S.K., Cledon, M., Sarma, S. and Verma, M., 2017. Biodegradation of polycyclic aromatic hydrocarbons (PAHs) by fungal enzymes: a review. *Journal of Environmental Sciences*, 51, pp. 52–74.

Karigar, C.S. and Rao, S.S., 2011. Role of microbial enzymes in the bioremediation of pollutants: a review. *Enzyme Research*, 2011, pp. 1–11.

Kumar, A, Singh, D, Kumar, R, Sakshi and Mahima, 2022a. Nutrient profile of vermicompost prepared from neem leaves. *Asian Journal of Microbiology, Biotechnology & Environmental Sciences*, 24(1), pp. 76–80.

Kumar, R., Gupta, R.K., Yadav, R., Saifi, R. and Yodha, K., 2022b. Eisenia fetida as protein source for growth enhancement of Heteropneustes fossilis. *Egyptian Journal of Aquatic Biology and Fisheries*, 26(2), pp. 577–588.

Kumar, R., Sharma, P., Gupta, R.K., Kumar, S., Sharma, M.M.M., Singh, S. and Pradhan, G., 2020. Earthworms for eco-friendly resource efficient agriculture. In: Kumar, S., Meena, R.S., Jhariya, M.K. (eds.), *Resources Use Efficiency in Agriculture* (pp. 47–84). Springer, Singapore.

Kumar, R. and Singh, D., 2022. Insecticide-tolerant bacterial population in Eisenia Fetida's gut and vermicast exposed to chlorantraniliprole and fipronil. *Applied Biological Research*, 24(3), pp. 273–279.

Kumar, R., Yadav, R., Gupta, R.K., Yodha, K., Kataria, S.K., Kadyan, P., Sharma, P. and Kaur, S., 2023. The earthworms: Charles Darwin's ecosystem Engineer. In: K.R. Hakeem (eds), *Organic Fertilizers* (pp. 1–7). IntechOpen, London.

Kumar, R., Yadav, R., Gupta, R.K., Pal, A., Yodha, K. and Kumar, A., 2024. Antioxidant enzyme activities and markers of oxidative stress in the life cycle of different Earthworm species. *Indian Journal of Experimental Biology*, 62(7), pp. 578–583.

Kumari, P., Alam, M. and Siddiqi, W.A., 2019. Usage of nanoparticles as adsorbents for waste water treatment: an emerging trend. *Sustainable Materials and Technologies*, 22, p. e00128.

Li, S., Jiang, Q., Liu, S., Zhang, Y., Tian, Y., Song, C., Wang, J., Zou, Y., Anderson, G.J., Han, J.Y. and Chang, Y., 2018. A DNA nanorobot functions as a cancer therapeutic in response to a molecular trigger in vivo. *Nature Biotechnology*, 36(3), pp. 258–264.

Ljubetič, A., Lapenta, F., Gradišar, H., Drobnak, I., Aupič, J., Strmšek, Ž., Lainšček, D., Hafner-Bratkovič, I., Majerle, A., Krivec, N. and Benčina, M., 2017. Design of coiled-coil protein-origami cages that self-assemble in vitro and in vivo. *Nature Biotechnology*, 35(11), pp. 1094–1101.

Mahanty, S., Chatterjee, S., Ghosh, S., Tudu, P., Gaine, T., Bakshi, M., Das, S., Das, P., Bhattacharyya, S., Bandyopadhyay, S. and Chaudhuri, P., 2020. Synergistic approach towards the sustainable management of heavy metals in wastewater using mycosynthe-sized iron oxide nanoparticles: biofabrication, adsorptive dynamics and chemometric modeling study. *Journal of Water Process Engineering*, 37, p. 101426.

Malik, J.A., Goyal, M.R. and Wani, K.A. (eds)., 2022. *Bioremediation and Phytoremediation Technologies in Sustainable Soil Management: Volume 4: Degradation of Pesticides and Polychlorinated Biphenyls*. CRC Press.

Masindi, V., Osman, M.S. and Tekere, M., 2021. Mechanisms and approaches for the removal of heavy metals from acid mine drainage and other industrial effluents. In: Inamuddin, Ahamed, M.I., and Lichtfouse, E. (eds.), *Water Pollution and Remediation: Heavy Metals* (pp. 513–537). Springer: Berlin/Heidelberg, Germany.

Mohamed, S.N., Thomas, N., Tamilmani, J., Boobalan, T., Matheswaran, M., Kalaichelvi, P., Alagarsamy, A. and Pugazhendhi, A., 2020. Bioelectricity generation using iron (II) molybdate nanocatalyst coated anode during treatment of sugar wastewater in microbial fuel cell. *Fuel*, 277, p. 118119.

Najafpoor, A., Norouzian-Ostad, R., Alidadi, H., Rohani-Bastami, T., Davoudi, M., Barjasteh-Askari, F. and Zanganeh, J., 2020. Effect of magnetic nanoparticles and silver-loaded magnetic nanoparticles on advanced wastewater treatment and disinfection. *Journal of Molecular Liquids*, 303, p. 112640.

Nogueira, H.P., Toma, S.H., Silveira, A.T., Carvalho, A.A., Fioroto, A.M. and Araki, K., 2019. Efficient Cr (VI) removal from wastewater by activated carbon superparamagnetic composites. *Microchemical Journal*, 149, p. 104025.

Noman, M., Shahid, M., Ahmed, T., Niazi, M.B.K., Hussain, S., Song, F. and Manzoor, I., 2020. Use of biogenic copper nanoparticles synthesized from a native Escherichia sp. as photocatalysts for azo dye degradation and treatment of textile effluents. *Environmental Pollution*, 257, p. 113514.

Priyadarshanee, M. and Das, S., 2021. Biosorption and removal of toxic heavy metals by metal tolerating bacteria for bioremediation of metal contamination: a comprehensive review. *Journal of Environmental Chemical Engineering*, 9(1), p. 104686.

Rebello, S., Sivaprasad, M.S., Anoopkumar, A.N., Jayakrishnan, L., Aneesh, E.M., Narisetty, V., Sindhu, R., Binod, P., Pugazhendhi, A. and Pandey, A., 2021. Cleaner technologies to combat heavy metal toxicity. *Journal of Environmental Management*, 296, p. 113231.

Ryu, H., Fuwad, A., Yoon, S., Jang, H., Lee, J.C., Kim, S.M. and Jeon, T.J., 2019. Biomimetic membranes with transmembrane proteins: state-of-the-art in transmembrane protein applications. *International Journal of Molecular Sciences*, 20(6), p. 1437.

Sharma, A., Thakur, M., Bhattacharya, M., Mandal, T. and Goswami, S., 2019. Commercial application of cellulose nano-composites – a review. *Biotechnology Reports*, 21, p. e00316.

Soares, P.R.S., Birolli, W.G., Ferreira, I.M. and Porto, A.L.M., 2021. Biodegradation pathway of the organophosphate pesticides chlorpyrifos, methyl parathion and profenofos by the marine-derived fungus Aspergillus sydowii CBMAI 935 and its potential for methylation reactions of phenolic compounds. *Marine Pollution Bulletin*, 166, p. 112185.

Takmil, F., Esmaeili, H., Mousavi, S.M. and Hashemi, S.A., 2020. Nano-magnetically modified activated carbon prepared by oak shell for treatment of wastewater containing fluoride ion. *Advanced Powder Technology*, 31(8), pp. 3236–3245.

Tripathi, M., Singh, D. N., Prasad, N., & Gaur, R. (2021). Advanced bioremediation strategies for mitigation of chromium and organics pollution in tannery. *Rhizobiont in Bioremediation of Hazardous Waste*, 195–215.

Tyagi, B. and Kumar, N., 2021. Bioremediation: principles and applications in environmental management. In: Saxena G., Kumar V. and Shah M. (eds.), *Bioremediation for Environmental Sustainability* (pp. 3–28). Elsevier, Netherland. https://www.sciencedirect.com/book/9780128205242/bioremediation-for-environmental-sustainability.

Vázquez-Núñez, E., Molina-Guerrero, C.E., Peña-Castro, J.M., Fernández-Luqueño, F. and de la Rosa-Álvarez, M.G., 2020. Use of nanotechnology for the bioremediation of contaminants: a review. *Processes*, 8(7), p. 826.

Verma, S. and Kuila, A., 2019. Bioremediation of heavy metals by microbial process. *Environmental Technology & Innovation*, 14, p. 100369.

Wang, M., Weiberg, A., Lin, F.M., Thomma, B.P., Huang, H.D. and Jin, H., 2016. Bidirectional cross-kingdom RNAi and fungal uptake of external RNAs confer plant protection. *Nature Plants*, 2(10), pp. 1–10.

Yadav, R., Gupta, R.K., Kumar, R. and Kaur, T., 2022b. Assessment of Toxicity of lead and nickel on the biochemical and immunological parameters of Earthworm, Eudrilus eugeniae. *Agricultural Science Digest*, D-5582, pp. 1–8.

Yadav, R., Gupta, R.K., Kumar, R., Kaur, T. and Moun, R., 2022a. Biochemical response of earthworm, Eisenia fetida to heavy metals toxicity. *Journal of Applied and Natural Science*, 14(3), pp. 990–998.

Yadav, R., Kumar, R., Gupta, R.K., Kaur, T., Yodha, K., Kour, A., Kaur, S. and Rajput, A., 2023. Heavy metal toxicity in earthworms and its environmental implications: a review. *Environmental Advances*, May 1, p. 100374.

2 Microbial Biofilms

Revolutionizing Fermentation and Bioremediation of Environmental Pollutants

Rajeeva Gaur, Saroj Bala,
Manikant Tripathi, and Ravi Kumar

2.1 INTRODUCTION

Biofilm is a microbial colony encapsulated or attached to the surface with the help of a polysaccharide having adhesive characteristics, which can then attach to a fixed matrix or float on the water surface. The chemical nature of biofilm is also variable because of the presence of different types of polysaccharides, lipids, or combinations of several other organic and inorganic components (Sauer et al., 2022). Biofilm may have some metallic compounds like calcium, magnesium, etc., along with some organic gelling compounds that may provide high temperature and cold resistance to microorganisms. Biofilm includes single or many types of microorganisms, such as bacteria, fungi, algae, and archaea, under varying environmental conditions (Schilcher and Horswill, 2020). A common example of biofilm is dental plaque in humans, which is a slimy layer of bacteria that forms on the surface of teeth and protects the microbes from high salt and sugar contents by reducing the water activity inside the gel for a longer period of time (Mishra et al., 2020).

Microbial biofilm has an important role in agriculture because it increases soil fertility and enhances plant growth. It is a beneficial structure that provides the possibility for easy transport of nutrients and genetic exchange to participating microorganisms, as well as protection from the surrounding environmental stresses. In nature, microbial biofilm also plays an important role in the degradation of many environmental pollutants and the production of various types of metabolites and products for human welfare (Figure 2.1). Most of these natural processes require the concerted effort of microbes with different metabolic capabilities, and it is likely that bacteria residing within biofilm communities carry out many of these complex processes (Karygianni et al., 2020).

The biofilm may tolerate very high pressure due to its elastic and high tensile strength, like biofilm produced from *Aureobasidium pullulans*. This versatile polysaccharide has many patents for its application, from contact lenses to biodegradable

DOI: 10.1201/9781003407317-2

17

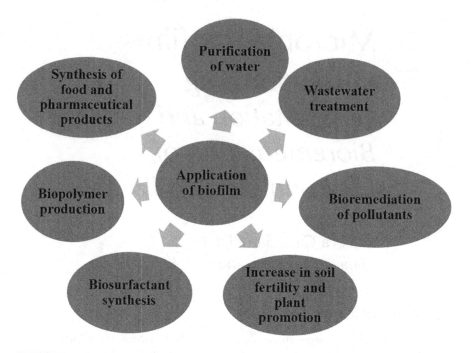

FIGURE 2.1   Beneficial role of biofilm in different areas.

plastic production. The biofilm for encapsulation provides firm attachment on porous solid surfaces that may be used in several metabolite productions using the immobilization of microbial cells (Pereira et al., 2021). Biofilm formation involves many steps (Figure 2.2). The steps of biofilm formation are as follows: (i) adhesion of planktonic cells to the support surface; (ii) formation of monolayers and cell proliferation; (iii) microcolonies formation; (iv) development of mature biofilm; and (v) detachment or dispersal of bacterial cells. The solid surface is the main habitat for various groups of microorganisms, which attach to the surface for longer periods of time to obtain nutrients from the surface as an abundance of nutrients is available. This phenomenon does occur for their long-term survival. The second approach is that microbes do not take nutrients from the attached solid substrate but rather from their colony formation, getting nutrients from the surrounding environment either through stagnant or flow hydrodynamic systems (Flemming et al., 2023). Therefore, under both conditions, biofilm supports their survival. It indicates that the microbes that are growing on the non-nutrient solid matrix stay for a longer period of time without being washed. Therefore, microbial numbers and activities are typically high on the surface. It is a normal tendency that microbial systems, whether aquatic, terrestrial, or always require surface area to adhere for nutrients, in which pili help bacteria. Bacteria form biofilm containing one, two, or several hundred different phylotypes (Ruhal and Kataria 2021). Therefore, biofilms are functional microbial communities or colonized microbes of friendly associates.

Biofilm growth is always more extensive and diverse than that of the microorganisms in the surroundings, which do not form biofilm. In a natural ecosystem, the

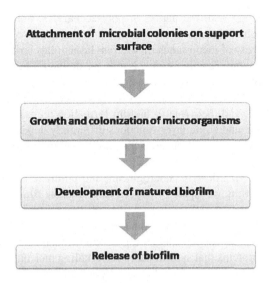

FIGURE 2.2   Steps of biofilm formation.

rate of biofilm formation by most of the organisms, viz., Azotobacter, Pseudomonas, Rhizobium, Arthrobacter, Bacillus, etc., is common and is able to produce various types of polysaccharides as well as form a film nearby the rhizosphere when appropriate water activity occurs. It results in their survival under film, where a variety of other microbes may exist in the nonporous film produced, which does not allow oxygen to enter the film, resulting in the survival of anaerobes to facultative aerobes on the surface itself (Gebreyohannes et al., 2019). The corresponding author has evaluated the old stagnant pond surface water, where algae were the dominant flora. The surface water also indicated the presence of aerobes and facultative anaerobes when tested through a nutrient agar stab using thioglycolate, showing an anaerobic count on the surface due to the support system for the survival of anaerobic bacteria because required gases and nutrients are available (Idrees et al., 2021). Different compounds like CO_2, NO_3, and other aliphatic acids may be required when the cells go deeper into the pond. Such conditions make the environment conducive for such microorganisms. Therefore, they may play a role in the recycling of nutrients, N_2 fixation, and other transformation processes at every segment of the pond. The science of biofilm chemistry and function needs more research. Biofilm potentials can be used as a model for processing various environmental pollutants, such as heavy metals, pesticides, and other toxicants, from water bodies, as well as a good barrier for groundwater pollution.

2.2   ARCHITECTURE AND PHYSIOLOGICAL FUNCTION OF BIOFILM

As we are aware, capsules and slime layers of bacteria may be thick, rigid, or flexible; permeable or nonpermeable; and short or big, depending on microorganisms and matrix size, along with the nature of polysaccharides. Some biofilms release dead and live microbial biomass as well as nutrients that are also passing in and out. Therefore, the porosity of membranes in different layers may vary (Campoccia et al., 2021). In this regard, the slime layer and capsule are defined. The slime layer is loose and dissolves easily, while capsules are typically firmly attached to the cell wall. The exact function and mechanism of transport of solutes, especially microbial metabolites within the film, and their role in biofilm formation are still questionable. The nature of the capsule and slime layer functions differently depending on the chemical configuration, which is still obscure. Extracellular polysaccharides of different natures play a key role in the development of biofilm, and different polysaccharides of microbial origin, viz., dextrin, scleroglucan, and pullulan, are being applied for the development of thin, transparent, oxygen-impermeable films useful for various applications (Fanesi et al., 2022).

Biofilm formation is an important signaling event to attach to the surface of solid support, which is governed by some signaling receptors that initiate biofilm-producing genes that encode proteins that synthesize intra-signaling molecules and initiate specific polysaccharides or other biomolecules depending on the nature of the habitat (abiotic and biotic). It may be one or more events, depending on the nutritional and physicochemical factors. Surface sensing is another important step in bacteria, which is perhaps governed by certain genes that initiate the formation of attachment segments by specific signaling processes for the secretion of several biomolecules (Figure 2.3). Such biomolecules could be able to form an effective semipermeable biomembrane of the desired value (Tang et al., 2022).

Some of the physiology and biochemistry of this have been beneficial, in that growth is triggered by the synthesis of cyclic dimeric monophosphate (c-di-GMP), a derivative of the nucleotide guanosine triphosphate. C-di-GMP is made by a series of proteins associated with membrane-integrated sensory proteins that in some way detect an opportunity for surface-associated growth (Kowalski et al., 2021). It is believed that c-di-GMP formation occurs by both triggering biofilm-specific gene expression and by altering enzymes in the cell that synthesize matrix material.

2.3 TREATMENT OF POLLUTANTS VIA BIOFILM: AN ECO-FRIENDLY APPROACH

Bioremediation is a process through which a specific group of microorganisms, mainly bacteria of the aerobic and anaerobic groups, molds, and yeast, have the capability of degrading xenobiotic compounds along with natural complex aromatic hydrocarbons of plant origin. Therefore, the main xenobiotic compounds are pesticides, insecticides, herbicides, lignin, petroleum hydrocarbons, and various other aromatic hydrocarbons present in the ecosystem. The main bacteria are *Pseudomonas, Arthrobacter, Xanthomonas, Bacillus*, etc., which can live in the biofilm (Guo et al., 2021). Therefore, a biofilm containing several genera and species of such bacteria can be large without losing their active biomass from the film, either in a fixed system or floating in a liquid system where they multiply and consume the xenobiotic compounds. The multilayer porous film provides different levels of

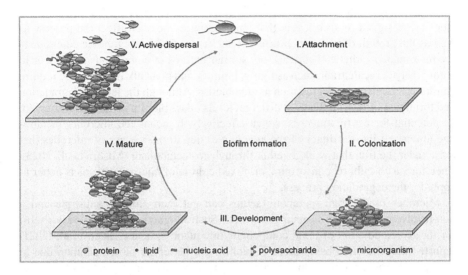

V. Active dispersal

I. Attachment

IV. Mature

Biofilm formation

II. Colonization

III. Development

○ protein　• lipid　∼ nucleic acid　⌇ polysaccharide　🦠 microorganism

FIGURE 2.3 Model showing the production of microbial biofilms (Yin et al., 2019).

transportation of toxic and nontoxic components, which makes the microbes comfortable and also supports the retention of enzymes responsible for the degradation of xenobiotic compounds through immobilized microbial cells and enzymes in the biofilm (Mammeri et al., 2019).

The current perspective of biofilm application for the remediation of toxic pollutants at the commercial level is one of the best alternatives. The biofilm architecture in which the role of Ca^{2+}, Mg^{2+}, and other metals may be associated with several cross-linking and layering of metals around the membranes through several modifications for the evaluation of biofilm efficacy is a matter of research in the bioremediation area (Ugya et al., 2023). The biofilm also manages the concentration of toxicants and the entrapment of specific molecules and enzymes for the degradation of xenobiotic compounds. The concentration of pollutants may also alter the permeability of membranes and shift microorganisms on a qualitative and quantitative basis. Bioremediation requires certain conditions, like temperature, oxygen requirements, and other physicochemical factors, which must be fulfilled by the biofilm for better efficiency. Such conditions also prevail in membrane bioreactors for bulk treatment of mixed types of chemical compounds. The film category and microbe types may be classified into aerobic and anaerobic treatment systems, where entirely two specific origins of microbes simultaneously work in different zones under the protection of film (Sathishkumar et al., 2023). The polluted ponds and lagoons of xenobiotic compounds are treated through this technology. The wastewater treated with biofilm systems has many benefits compared to free cells, such as operational flexibility, stability to environmental changes, and enhanced ability to degrade recalcitrant compounds using microbial enzymes.

A biofilm is a group of microorganisms attached to a solid support or biological or inert surface and encased in a self-synthesized matrix comprising water, proteins, carbohydrates, and extracellular DNA. Biofilm-mediated remediation is an eco-friendly and feasible alternative for treating environmental toxicants. The biofilm characteristics are designed in such a way that ionic strength does not affect the process as well as the growth of microorganisms (Mishra et al., 2023). Xenobiotic compounds are man-made synthetic chemicals, but similar natures of compounds also occur in nature that are recalcitrant, such as lignin, tannins, and plant alkaloids, which require similar mechanisms of degradation as xenobiotics. Although the level of degradation and formation of intermediates at different levels under biofilm is not well discussed, its potential for net treatment is shown effectively. It means the microbes metabolize almost all intermediates of the xenobiotics due to the variety of microbes that exist under the film that work together through co-metabolism (Mishra et al., 2023). Therefore, a co-culture, consortium, or mixed culture of microorganisms is better to apply for the degradation process.

Microbes of different metabolic setups can get their suitable substrate and a non-involvement environment to grow well. Such interactions will be very useful for the bioremediation of toxic compounds in aquatic systems. Balan et al. (2021) reported that the microbes present in biofilm are better for bioremediation due to their greater tolerance to pollutants and various environmental stresses that degrade several toxic pollutants through different metabolic pathways. In biofilm, microbes are immobilized in a self-synthesized matrix that provides protection from stresses, contaminants, and predatory animals.

In natural conditions, microbes with the capability of producing biofilm of different characteristics may also synthesize for the development of new combinations of biofilm, which may retain a variety of microorganisms for long-term existence, leading to the fast degradation of xenobiotic compounds. Some microbes that do not form film or do not live in the film may also support biofilm formation, which should be identified for the development of new microbial associations for effective bioremediation (Murshid et al., 2023). In a natural ecosystem, it has been identified that some microbes that are friendly exist with certain microbes that are specific and need a higher number for better biofilm development. Moreover, it is necessary to identify the biofilm structure and friendly microbial associates for the natural treatment process. The three-dimensional structure of exopolysaccharides (EPS) with reduced oxygen concentrations promotes faster degradation of many types of pollutants in the environment. Several researchers reported that the EPS from cyanobacteria acts as a biosorbent and remediates heavy metals from the aqueous environment.

The microbial world of biofilm is either floating or attached to the next solid surface inside the biofilm. The microbial association of biofilm may exist with various interactions, viz., symbiosis, commensalism, predation, amensalism, competition, etc. Such activities help in the development of selective microorganisms as well as the degradation mechanisms of pollutants in the environment (Zinicovscaia and Balintova, 2023). It means there are some members who have the capability to form various types of polysaccharides of different chemical natures, leading to different types of biofilm. Their friendly associates live together in different numbers, helping them to make favorable conditions. Sometimes the entire microbial community is developed within the polysaccharide matrix. A group of researchers reported that the biofilm formation in microorganisms is a chemotaxis-dependent response, and flagellar-dependent motility such as swimming, swarming, twitching, and quorum sensing in the presence of xenobiotics initiates the microbial activity for the formation of several types of biofilms in soil and water conditions to assist microbes in coordinating movement toward pollutant and improved biodegradation (Liaqat et al., 2023; Ahsan et al., 2023). Biofilm is increasingly used as an indicator for monitoring and evaluating contamination in water systems by assessing its properties, such as change in biomass, species composition, pigment production, photosynthesis, and enzymatic activity. Many researchers reported the change in biofilm community or alteration in pigment composition of microbial biofilm upon exposure to environmental pollutants such as heavy metals and suggested biofilm as an environmental pollution indicator (Syed et al., 2023). The role of biofilm in food, medicine, and agriculture too is important, but the science of biofilm is almost the same, with varying strains of microbes used for environmental sanitation and several problems associated with biofilm.

2.4 MICROBIAL ENZYMES UNDER BIOFILM

The microbes are specific and produce certain enzymes responsible for the degradation of xenobiotic compounds. The main enzymes are peroxidase, phenoxidase, mono- and di-oxygenases, laccases, etc. These enzymes are required for the degradation of aromatic hydrocarbons, utilizing various metabolic pathways like beta-ketoadipic, caprolate, and mandalate pathways that have certain intermediates. Microbial enzymes are primarily responsible for the degradation of sewage and other industrial wastewater, i.e., pectinases, amylases, chitinases, cellulases, xylanases, proteases, lipases, phosphatidases, dihydrogenases, mono- and di-oxygenases, esterases, peroxidases, laccases, etc. (Ramakrishnan et al., 2022). These enzymes are frequently produced by several groups of microorganisms, mainly bacteria, yeast, and microfungi of aquatic origin, under the biofilm and work just like trickling filters for the treatment of industrial effluents. Several industrial effluents contain melanoidins, a recalcitrant color compound of distillery effluent that requires specific bacteria like *Pseudomonas, Acetobacter, Bacillus, Arthrobacter,* and *Xanthomonas,* and several yeasts like *Candida* and *Torula* for effective treatment. Some of the actinomycetes, like *Actinoplanes, Thermomonospora, Nocardia*, and *Streptomyces*, form active biofilm and may retain biofilm and do the degradation of toxic compounds, although the mechanism of survival and entrapment/encapsulation or any other mechanism of the retention of cells or enzymes is still a matter of research that may explore its application using a suitable matrix and design of biofilm for the treatment of many toxic compounds in a bioreactor (Sonawane et al., 2022). Enzyme-linked biofilm characteristics are also very important because biotransformation of organic matter and leaching of metals can be best remediated by biofilm application at the natural and bioreactor levels. The binding of metals is governed by a variety of microbial enzymes, which catalyze bioconversion and provide the molecular basis for the metabolic pathways essential for the growth of living cells. Several biofilm reactor designs have been proposed by several engineers and biotechnologists using multistage and multiprocess systems using solid and semisolid states based on the nature of pollutants for effective treatment of the most selective and toxic components of effluent because biofilm has selective permeability (Ajijah et al., 2023). Such technology can also be modified according to the nature of microbes as well as the types of effluent.

2.5   APPLICATION OF BIOFILM IN FERMENTATION TECHNOLOGY

Biofilm has several benefits in the fermentation industry due to the existence of long-term biological activity that achieves continuous production and treatment processes of many value-added bioproducts economically. Biofilm reactors are more cost-effective than batch processes of free cells because of the reduction in reactor preparation, cell growth, and product recovery (Martyniuk et al., 2022). Biofilms of bacteria, fungi, and their enzymatic products can act as biocatalysts to provide high production. Many microbes in the environment can exist in the form of biofilms, such as *Bacillus subtilis, Escherichia coli, Clostridium acetobutylicum, Saccharomyces cerevisiae*, etc. Various processes using biofilm have been employed for commercial production over the past decades. Biofilm has been used in the food sector for producing organic acids, polysaccharides, ethanol, butanol, etc. Biofilm provides favorable conditions for the growth of cells during the industrial fermentation process. Some researchers reported efficient biofilm-based fermentation strategies for L-proline and L-threonine production using *Corynebacterium gluctamicum* and *Escherichia coli*, respectively (Thallinger et al., 2013; Sharahi et al., 2019).

The role of biofilm in the fermentation industry occurs with the immobilization of microbial cells. The technique can be applied only when microbes have the ability to form microcolony on solid surfaces, which means film attachment to a solid surface because all microbial systems do not have such characteristics. Immobilization of microbial cells on suitable solid materials like alumina, polyurethane, and several others, depending on the nature of microorganisms, is frequently used in the fermentation industry. Actually, this technique provides a continuous mode of fermentation of microbial metabolites, leading to an economic metabolite yield over normal continuous fermentation (Wang and Chen, 2009; Kaur et al., 2023). Generally, continuous fermentation is a submerged system that obeys the principles of chemo and turbidostatic culture, but during processing, the microbial biomass flows in a higher amount, and in the normal course, the biomass in the initial fermentation is reduced, which is then recycled through a large centrifuge automated with biomass indicators. The running cost of several large centrifuges is very high, which increases the cost of metabolites. However, this process is economically over batch processes where culture production in bulk involves energy, cost, time, and contamination. Therefore, immobilization in the fermentation process does not require the recycling of biomass, as biomass is always formed in the main fermenter due to the firm attachment of microbes to its suitable solid support. Tatsaporn and Kornkanok (2020) studied ethanol production by the fermentation of rice bran hydrolysate using *Zygomonas mobilis* biofilm immobilization on DEAE-cellulose. This microcolony formation on the solid support system multiplies vertically, detaching the young cells from the medium. The original culture is attached firmly in live form, with continuous multiplication generating young live cultures. Thus, this process is always most economical and less time-consuming with minimum operation parameters. However, during screening, if microbes have the characteristics to attach through specific polysaccharide films and fix with the original colony on the solid surface, then immobilization of microbial cells will be feasible for the bioprocess system.

Further, secondary microbial metabolites always prefer batch fermentation processes, while the application of biofilm can make the process possible in a continuous system, which is the most effective and economical process (Jiang et al., 2021). The fermentation by free cells would increase the cost of production, and the free cells are also challenged by stress conditions such as shear forces during fermentation, resulting in decreased cell viability. Such conditions require a specific solution to improve the fermentation process. Biofilm-based immobilized fermentation is better compared to free-cell fermentation owing to its advantages, such as protection by the biofilm matrix, enhanced metabolic activities, and repeated use of immobilized cells of biofilm. Several researchers employed the biofilm of microorganisms such as *Escherichia coli*, *Clostridium acetobutylicum*, and *Corynebacterium glutamicum* in immobilized batch or continuous fermentation (Lacroix and Yildirim, 2007; Tripathi et al., 2023). Most of the microbial metabolite production or treatment of effluents require specific microbial communities in different segments because they produce specific enzymes or metabolites responsible for production or degradation. In such a process, the multiple reactions and substrates are transformed in a single bioreactor; thus, biofilm technology can only achieve such performance in a cost-effective manner. The biofilm technology also checks the microbial contaminants during fermentation, which adds quality products in a cost-effective manner (Table 2.1). For a clearer understanding, immobilization may be divided into two types based on the process and mechanisms adopted with physical and chemical principles directly with active and passive immobilization of cells.

2.5.1 ACTIVE IMMOBILIZATION

Active immobilization is the entrapment and binding of cells by physical and chemical forces. Physical entrapment within porous materials is the most widely used method of cell immobilization. Various matrices are being used for the immobilization of cells. The main porous polymers are agar, alginate, k-carrageenan, polyacrylamide, chitosan, gelatin, porous metal screens, polyurethane, silica gel, polystyrene, etc. (Al-Amshawee et al., 2020). In this polymer, the beads should be porous enough to allow the substrates and products in and out of the beads. Therefore, the cells and support matrix are prepared by different methods like precipitation of polymers, gelation, ion exchange, polymerization, polycondensation, etc. Therefore, different principles of adsorption, absorption, entrapment, and membrane-based entrapment are based on porosity and several others like covalent binding, etc. (Baidamshina et al., 2020).

2.5.2 PASSIVE IMMOBILIZATION

Passive immobilization is facilitated by biological films, in which multilayer growth of cells occurs on a solid support surface. The support material may be inert or biologically active and fixed or floating in a submerged solid substrate level (Liang et al., 2020). The interaction of living cells to support material by specific binding

TABLE 2.1

Biofilm Application in Fermentation Technology

S.No.	Application	Remarks	Reference
1.	Enhanced bioproduction	Immobilizing microorganisms in biofilms improves production rates and stability, particularly in the production of biofuels and enzymes.	Gunes (2021); Baidamshina et al. (2020); Jiang et al. (2021)
2.	Bioremediation	Biofilm reactors treat wastewater contaminated with pollutants, enhancing microbial activity for more efficient pollutant degradation.	
3.	Improved product quality	Biofilms enable precise control of microbial growth and metabolite production, ensuring better quality and consistency in products like pharmaceuticals and bioplastics.	
4.	Biopharmaceuticals	Biofilm-based fermentation increases cell densities, reducing production volumes and simplifying downstream processing in biopharmaceutical production.	
5.	Sustainable agriculture	Biofilm-coated seeds and plant roots promote beneficial microbial interactions, enhancing nutrient uptake and disease resistance in crops.	
6.	Biofilm sensors	Biofilms can serve as sensors to monitor environmental conditions and microbial activity, providing real-time insights for process optimization.	

agents and different forces for such a process is again a matter of research. In mixed microbial films, the presence of some polymer-producing organisms is capable of forming biofilm and enhances the stability of the biofilm. The thickness of biofilm depends on the types, microbial load, nature, and quantity of pollutants (Ünal Turhan et al., 2019; Pathak et al., 2022). The thickness and size may be reduced or increased depending on the requirements of microorganisms and environmental conditions. A microenvironment inside a thick biofilm is created with either one or different modes of physiology. Biofilm formation is just for the long-term survival and maintenance of the population of a particular niche in the ecosystem, which depends on the availability of nutrients in a particular situation (Mahdinia and Demirci, 2020).

In a biofilm, nutrients diffuse in it and the microbial product diffuses out into a liquid medium. The nutrient and product profiles in the film are important factors affecting cellular physiology and metabolism. Biofilm cultures have almost the same advantages as those of immobilized cell systems in suspension culture. The rate of metabolite formation and biomass, along with substrate concentration, is a set process that may achieve continuous fermentation (Bastarrachea et al., 2022).

2.6 MICROSCOPIC EXAMINATION OF BIOFILM

Biofilm microbial systems can be effectively studied with the help of confocal scanning laser microscopy. It is a computerized microscope that couples a laser source with a light microscope. This generates a three-dimensional digital image of the microorganisms and biological specimens in the sample. The laser beam is adjusted in such a way that only a particular layer within a specimen is visible (Reichhardt and Parsek, 2019). By precisely illuminating only a single plane of focus, stray light from other focal planes is eliminated. Thus, when diverting a relatively thick specimen of biofilm cells under layers, it can be seen by adjusting the laser beam (Liu et al., 2021). By illuminating a specimen with a laser whose intensity varies as a wave, it has been possible to improve the 0.2 μm resolution of the compound light microscope to a limit of about 0.1 μm. To evaluate the properties of the biofilm, fluorescent dyes are used to make the microorganisms more distinct. But without dye, it can also be seen by adjusting the laser beam. Microscopy is generally used in the study of microbial ecology, especially for identifying phylogenetically distinct populations of cells present in a microbial habitat or for resolving different components of biofilms (Flemming et al., 2021; Nirwati et al., 2019).

2.7 HYPOTHETICAL MODELS OF BIOFILM

The microbes that produce a higher amount of polysaccharide, along with other biopolymers having specific lipids and proteins, may provide specific strength and sustainability toward high- or low-temperature permeability to metallic ions (Bhowmik et al., 2022). The carbohydrate polymers such as pullulan, dextrin, and alginate provide different functions; for example, pullulan produces high tensile strength and an oxygen-impermeable nature to the film, while dextrin provides selective translocation of compounds and alginates help in the entrapment of bacterial cells, supporting a higher rate of immobilization of microbial cells in the biofilm matrix (Barani et al., 2022). These polysaccharides may be polymerized by other groups of inorganic and organic compounds available in the ecosystem, supporting a large surface area binding capacity and cross-linking to the polymer by increasing the enzyme activity for higher transformation principles supported by specific groups of bacteria, fungi, and yeast, depending on the architecture of the biofilm. It has been observed that *Pseudomonas, Xanthomonas,* and *Arthrobacter* are the general bacteria and *Candida* and *Endomycopsis* are common yeasts isolated from several biofilm materials of the natural ecosystem (Willett et al., 2019). The selectivity of biofilm is very diverse in ecosystems that have different capabilities for the transformation of compounds along with the mobility of toxic components inside the film as well as remedial measures because several mechanisms are involved in the transformation process depending on the nature of enzymes and their permeability levels (de Freitas et al., 2021).

2.8   BIOFILM IN MICROBIAL METABOLITES PRODUCTION

Membrane film bioreactors are always better for some metabolites, such as hormones, enzymes, vaccines, or specific proteins and organic compounds, which are affected by end-product inhibition; therefore, higher product concentrations do not affect the fermentation process because the metabolite is separated during the course of production to a satisfactory level of inhibition (Crabbé et al., 2019). In many cases, ethanol and other alcoholic components can also be separated during the fermentation to minimize the inhibition levels of microbes; therefore, continuous production of such metabolites can be achieved through such biofilm, reducing the cost of downstream processing and also helping in batch and continuous fermentation (Table 2.2). Further, the biofilm formation in a specialized column with different dimensions and regular movements in the bioreactor at different agitation rates may improve the efficiency of the membrane bioreactor (Wu et al., 2023).

Biofilm produced by various microorganisms has a very diverse architecture depending on the requirements of particular microbial groups, in which micronutrient exchange, toxic metabolites, and other protective or inhibitive substances are

TABLE 2.2
Advantages and Challenges of Biofilm-Based Fermentation

S.No.	Aspects	Advantages	Challenges	Reference
1.	Enhanced microbial growth	1. Higher cell density and productivity. 2. Improved resistance to adverse conditions.	1. Biofilm formation can be unpredictable. 2. Difficulty in controlling biofilm growth.	Lü et al. (2023); Lee et al. (2022); Sun et al. (2023); Sinharoy et al. (2022)
2.	Nutrient utilization	1. Efficient nutrient utilization. 2. Reduced nutrient waste.	1. Risk of nutrient limitation in the core. 2. Formation of dead zones in the biofilm.	
3.	Biofilm protection	1. Enhanced resistance to toxins. 2. Protection against predation.	1. Difficulty in monitoring biofilm health. 2. Risk of biofilm contamination.	
4.	Enhanced metabolic pathways	1. Potential for diverse metabolic activities. 2. Synthesis of valuable compounds.	1. Complex metabolic interactions. 2. Potential for competition among species.	
5.	Biofilm engineering	1. Scalability and ease of immobilization. 2. Reduced downstream processing costs.	1. Technical challenges in reactor design. 2. Maintaining stable biofilm architecture.	
6.	Biofilm-based productivity	1. High product yields. 2. Reduced fermentation time.	1. Risk of biofilm detachment. 2. Limited knowledge of biofilm dynamics.	

selectively in or out of semipermeable biofilm. The lipids and proteins associated with biofilm just behave like the membranes of the living system. The translocation of specific solutes and metabolites from the biofilm to the existing system is facilitated by a highly selective automated barrier. The exact configuration of several types of biofilms chemically and structurally varies as per the requirements of the environment and changing solute and solution levels (Coenye et al., 2020). Such an aspect is a matter of research for the use of biofilm synthesized by microorganisms of different associations. Such work may solve several facts and improve the management of microorganisms in the areas of agriculture, environment, industry, and medical microbiology, either for their effective use in the treatment of pollutants or to manage plant growth-promoting microbes in agriculture or in industrial fermentation processes, as well as in the control of plant and animal diseases. Further, the biomaterials for the design of thermotolerance, salt/nutrient tolerance, and exchange through biofilm may persist for a longer period for the exchange of specific components from biofilm on an in/out basis without affecting the surrounding microbes. A detailed account accounts for the exchange of organic and inorganic chemicals along with selective barriers at changing temperature, pH, and water activity (Ćirić et al., 2019). These may also be studied in detail using sophisticated ultrasonication techniques regarding biofilm chemistry, architecture, and design for effective use in various areas. Very scanty information in this regard is available in these aspects. Therefore, there is a need to understand the membrane chemistry, translocation, and exchange of various metabolites. The study of the biofilm chemistry of polluted or freshwater pond systems and the rhizosphere of specific plants should be evaluated. In this review, the authors have tried to explore the potential of biofilm, especially for the treatment of pollutants, and its role in fermentation.

2.9   CONCLUSIONS

Microbial biofilm is a new emerging area for researchers to work in the fields of environment, industry, agriculture, and health. This science enhances the proliferation and colonization of several microbes on the surface and protects cells in an adverse environment. The potential of microbes surrounded by biofilm has recently been realized for bioremediation in sustainable environments and industries. The beneficial interactions of microorganisms in biofilm attract attention to pollutant degradation in a large-scale bioreactor. Moreover, biofilm is also an interesting area in fermentation technology where it can be used for many beneficial purposes via immobilization processes and reduce microbial contamination through different permeability levels of the film. Therefore, a better understanding of the role of microbial biofilm and its applications in various fields is necessary to develop new strategies for effective bioremediation and fermentation technology.

REFERENCES

Al-Amshawee, S., Yunus, M.Y.B.M., Vo, DV.N. et al. 2020. Biocarriers for biofilm immobilization in wastewater treatments: a review. *Environmental Chemistry Letters.* 18, pp. 1925–1945. https://doi.org/10.1007/s10311-020-01049-y

Ahsan, Y., Qurashi, A.W., Liaqat, I., Latif, A., Afzaal, M., Khalid, A. and Sardar, A.A.. 2023. "Efficacy of rice husk and halophilic bacterial biofilms in the treatment of saline water." *Journal of Basic Microbiology*. 63(8), pp. 855–867. doi: 10.1002/jobm.202300062.

Ajijah, N., Fiodor, A., Pandey, A.K., Rana, A. and Pranaw, K. 2023. "Plant growth-promoting bacteria (PGPB) with biofilm-forming ability: a multifaceted agent for sustainable agriculture." *Diversity*, 15(1), p. 112.

Baidamshina, D.R., Koroleva, V.A., Trizna, E.Y., Pankova, S.M., Agafonova, M.N., Chirkova, M.N., Vasileva, O.S., Akhmetov, N., Shubina, V.V., Porfiryev, A.G. and Semenova, E.V. 2020. "Anti-biofilm and wound-healing activity of chitosan-immobilized ficin." *International Journal of Biological Macromolecules*, 164, pp. 4205–4217.

Balan, B., Dhaulaniya, A.S., Varma, D.A., Sodhi, K.K., Kumar, M., Tiwari, M. and Singh, D.K. 2021. "Microbial biofilm ecology, in silico study of quorum sensing receptor-ligand interactions and biofilm mediated bioremediation." *Archives of Microbiology*, 203, pp. 13–30.

Barani, N., Sarabandi, K., Kotov, N.A., Vanepps, J.S., Elvati, P., Wang, Y. and Violi, A. 2022. "A multiphysics modeling of electromagnetic signaling phenomena at kHz-GHz frequencies in bacterial biofilms." *IEEE Access*, 10, pp. 39344–39361.

Bastarrachea, L.J., Britt, D.W. and Demirci, A. 2022. "Development of bioactive solid support for immobilized Lactococcus lactis biofilms in bioreactors for the production of nisin." *Food and Bioprocess Technology*, 15(1), pp. 132–143.

Bhowmik, P., Rajagopal, S., Hmar, R.V., Singh, P., Saxena, P., Amar, P., Thomas, T., Ravishankar, R., Nagaraj, S., Katagihallimath, N. and Sarangapani, R.K. 2022. "Validated in silico model for biofilm formation in Escherichia coli." *ACS Synthetic Biology*, 11(2), pp. 713–731.

Campoccia, D., Montanaro, L. and Arciola, C.R. 2021. "Extracellular DNA (eDNA). A major ubiquitous element of the bacterial biofilm architecture." *International Journal of Molecular Sciences*, 22(16), p. 9100.

Ćirić, A.D., Petrović, J.D., Glamočlija, J.M., Smiljković, M.S., Nikolić, M.M., Stojković, D.S. and Soković, M.D. 2019. "Natural products as biofilm formation antagonists and regulators of quorum sensing functions: a comprehensive review update and future trends." *South African Journal of Botany*, 120, pp. 65–80.

Coenye, T., Kjellerup, B., Stoodley, P. and Bjarnsholt, T. 2020. "The future of biofilm research– Report on the '2019 Biofilm Bash." *Biofilm*, 2, p. 100012.

Crabbé, A., Jensen, P.Ø., Bjarnsholt, T. and Coenye, T. 2019. "Antimicrobial tolerance and metabolic adaptations in microbial biofilms." *Trends in Microbiology*, 27(10), pp. 850–863.

de Freitas, S.B., Wozeak, D.R., Neto, A.S., Cardoso, T.L. and Hartwig, D.D. 2021. "A hypothetical adhesin protein induces anti-biofilm antibodies against multi-drug resistant Acinetobacter baumannii." *Microbial Pathogenesis*, 159, p. 105112.

Fanesi, A., Martin, T., Breton, C., Bernard, O., Briandet, R. and Lopes, F. 2022. "The architecture and metabolic traits of monospecific photosynthetic biofilms studied in a custom flow-through system." *Biotechnology and Bioengineering*, 119(9), pp. 2459–2470.

Flemming, H.C., Baveye, P., Neu, T.R., Stoodley, P., Szewzyk, U., Wingender, J. and Wuertz, S. 2021. "Who put the film in biofilm? The migration of a term from wastewater engineering to medicine and beyond." *NPJ Biofilms and Microbiomes*, 7(1), p. 10.

Flemming, H.C., van Hullebusch, E.D., Neu, T.R., Nielsen, P.H., Seviour, T., Stoodley, P., Wingender, J. and Wuertz, S. 2023. "The biofilm matrix: multitasking in a shared space." *Nature Reviews Microbiology*, 21(2), pp. 70–86.

Gebreyohannes, G., Nyerere, A., Bii, C. and Sbhatu, D.B. 2019. "Challenges of intervention, treatment, and antibiotic resistance of biofilm-forming microorganisms." *Heliyon*, 5(8), e02192. doi: 10.1016/j.heliyon.2019.e02192.

Gunes, B. 2021. "A critical review on biofilm-based reactor systems for enhanced syngas fermentation processes." *Renewable and Sustainable Energy Reviews*, 143, p. 110950.

Guo, R., Li, K., Tian, B., Wang, C., Chen, X., Jiang, X., He, H. and Hong, W. 2021. "Elaboration on the architecture of pH-sensitive surface charge-adaptive micelles with enhanced penetration and bactericidal activity in biofilms." *Journal of Nanobiotechnology*, 19(1), pp. 1–18.

Idrees, M., Sawant, S., Karodia, N. and Rahman, A. 2021. "Staphylococcus aureus biofilm: morphology, genetics, pathogenesis and treatment strategies." *International Journal of Environmental Research and Public Health*, 18(14), p. 7602.

Jiang, Y., Liu, Y., Zhang, X., Gao, H., Mou, L., Wu, M., Zhang, W., Xin, F. and Jiang, M. 2021. "Biofilm application in the microbial biochemicals production process." *Biotechnology Advances*, 48, p. 107724.

Karygianni, L., Ren, Z., Koo, H. and Thurnheer, T. 2020. "Biofilm matrixome: extracellular components in structured microbial communities." *Trends in Microbiology*, 28(8), pp. 668–681.

Kaur, M., Bhatia, S., Gupta, U., Decker, E., Tak, Y., Bali, M., Gupta, V.K., Dar, R.A. and Bala, S. 2023. "Microalgal bioactive metabolites as promising implements in nutraceuticals and pharmaceuticals: inspiring therapy for health benefits." *Phytochemistry Reviews*, pp. 1–31. doi: 10.1007/s11101-022-09848-7.

Kowalski, C.H., Morelli, K.A., Stajich, J.E., Nadell, C.D. and Cramer, R.A. 2021. "A heterogeneously expressed gene family modulates the biofilm architecture and hypoxic growth of Aspergillus fumigatus." *MBio*, 12(1), pp. 10–1128.

Lacroix, C. and Yildirim, S. 2007. "Fermentation technologies for the production of probiotics with high viability and functionality." *Current Opinion in Biotechnology*, 18(2), pp. 176–183.

Lee, H.S., Xin, W., Katakojwala, R., Mohan, S.V. and Tabish, N.M. 2022. "Microbial electrolysis cells for the production of biohydrogen in dark fermentation-A review." *Bioresource Technology*, 363, 127934.

Liang, C., Ding, S., Sun, W., Liu, L., Zhao, W., Zhang, D., Ying, H., Liu, D. and Chen, Y. 2020. "Biofilm-based fermentation: a novel immobilisation strategy for Saccharomyces cerevisiae cell cycle progression during ethanol production." *Applied Microbiology and Biotechnology*, 104, pp. 7495–7505.

Liaqat, I., Muhammad, N., Ara, C., Hanif, U., Andleeb, S., Arshad, M., Aftab, M.N., Raza, C. and Mubin, M. 2023. "Bioremediation of heavy metals polluted environment and decolourization of black liquor using microbial biofilms." *Molecular Biology Reports*, 50(5), pp. 3985–3997.

Liu, Y., Wu, L., Han, J., Dong, P., Luo, X., Zhang, Y. and Zhu, L. 2021. "Inhibition of biofilm formation and related gene expression of Listeria monocytogenes in response to four natural antimicrobial compounds and sodium hypochlorite." *Frontiers in Microbiology*, 11, p. 617473.

Lü, J., Ren, G., Hu, Q., Rensing, C. and Zhou, S. 2023. "Microbial biofilm-based hydrovoltaic technology." *Trends in Biotechnology*, 41(9), pp. 1155–1167.

Mahdinia, E. and Demirci, A. 2020. "Biofilms in fermentation for the production of value-added products." In: Abdul, B. (ed.), *Microbial Biofilms* (pp. 73–108). CRC Press, Boca Raton, FL.

Mammeri, N.E., Hierrezuelo, J., Tolchard, J., Cámara-Almirón, J., Caro-Astorga, J., Álvarez-Mena, A., Dutour, A., Berbon, M., Shenoy, J., Morvan, E. and Grélard, A. 2019. "Molecular architecture of bacterial amyloids in Bacillus biofilms." *FASEB Journal*, 33(11), pp. 12146–12163.

Martyniuk, N., Souza, M.S., Bastidas Navarro, M., Balseiro, E. and Modenutti, B. 2022. "Nutrient limitation affects biofilm enzymatic activities in a glacier-fed river." *Hydrobiologia*, 849(13), pp. 2877–2894.

Mishra, R., Panda, A.K., De Mandal, S., Shakeel, M., Bisht, S.S. and Khan, J. 2020. Natural anti-biofilm agents: strategies to control biofilm-forming pathogens." *Frontiers in Microbiology*, 11, p. 566325.

Mishra, V., Mudgal, N., Rawat, D., Poria, P., Mukherjee, P., Sharma, U., Kumria, P., Pani, B., Singh, M., Yadav, A. and Farooqi, F. 2023. "Integrating microalgae into textile wastewater treatment processes: advancements and opportunities." *Journal of Water Process Engineering*, 55, p. 104128.

Murshid, S., Antonysamy, A., Dhakshinamoorthy, G., Jayaseelan, A. and Arivalagan, P. 2023. "A review on biofilm-based reactors for wastewater treatment: recent advancements in biofilm carriers, kinetics, reactors, economics, and future perspectives." *Science of the Total Environment*, 892, p. 164796.

Nirwati, H., Sinanjung, K., Fahrunissa, F., Wijaya, F., Napitupulu, S., Hati, V.P., Hakim, M.S., Meliala, A., Aman, A.T. and Nuryastuti, T. 2019. "Biofilm formation and antibiotic resistance of Klebsiella pneumoniae isolated from clinical samples in a tertiary care hospital, Klaten, Indonesia." *BMC Proceedings*, 13(11), pp. 1–8.

Pathak, N., Singh, S., Singh, P., Singh, P.K., Singh, R., Bala, S., Thirumalesh, B.V., Gaur, R. and Tripathi, M. 2022. "Valorization of jackfruit waste into value added products and their potential applications." *Frontiers in Nutrition*, 9, p. 1061098.

Pereira, R., dos Santos Fontenelle, R.O., De Brito, E.H.S. and De Morais, S.M. 2021. "Biofilm of Candida albicans: formation, regulation and resistance." *Journal of Applied Microbiology*, 131(1), pp. 11–22.

Ramakrishnan, R., Singh, A.K., Singh, S., Chakravortty, D. and Das, D. 2022. "Enzymatic dispersion of biofilms: an emerging biocatalytic avenue to combat biofilm-mediated microbial infections." *Journal of Biological Chemistry*, 298, p. 102352.

Reichhardt, C. and Parsek, M.R. 2019. "Confocal laser scanning microscopy for analysis of Pseudomonas aeruginosa biofilm architecture and matrix localization." *Frontiers in Microbiology*, 10, p. 677.

Ruhal, R. and Kataria, R. 2021. "Biofilm patterns in gram-positive and gram-negative bacteria." *Microbiological Research*, 251, p. 126829.

Sathishkumar, K., Naraginti, S., Lavanya, K., Zhang, F., Ayyamperumal, R. and Liu, X. 2023. "Intimate coupling of gC3N4/CdS semiconductor on eco-friendly biocarrier loofah sponge for enhanced detoxification of ciprofloxacin." *Environmental Research*, 235, p. 116558.

Sauer, K., Stoodley, P., Goeres, D.M., Hall-Stoodley, L., Burmølle, M., Stewart, P.S. and Bjarnsholt, T. 2022. "The biofilm life cycle: expanding the conceptual model of biofilm formation." *Nature Reviews Microbiology*, 20(10), pp. 608–620.

Schilcher, K. and Horswill, A.R. 2020. "Staphylococcal biofilm development: structure, regulation, and treatment strategies." *Microbiology and Molecular Biology Reviews*, 84(3), pp. 10–1128.

Sharahi, J.Y., Azimi, T., Shariati, A., Safari, H., Tehrani, M.K. and Hashemi, A. 2019. "Advanced strategies for combating bacterial biofilms." *Journal of Cellular Physiology*, 234(9), pp. 14689–14708.

Sinharoy, A., Pakshirajan, K., Lens, P.N.L. 2022. "Syngas Fermentation for Bioenergy Production: Advances in Bioreactor Systems." In: Sinharoy, A., and Lens, P.N.L. (eds.) Renewable Energy Technologies for Energy Efficient Sustainable Development. *Applied Environmental Science and Engineering for a Sustainable Future*. Springer Cham, Springer Nature Switzerland. (pp.325–358).

Sonawane, J.M., Rai, A.K., Sharma, M., Tripathi, M. and Prasad, R. 2022. "Microbial biofilms: recent advances and progress in environmental bioremediation." *Science of the Total Environment*, 824, p. 153843.

Sun, H., Xiao, K., Zeng, Z., Yang, B., Duan, H., Zhao, H. and Zhang, Y. 2022. "Electroactive biofilm-based sensor for volatile fatty acids monitoring: a review." *Chemical Engineering Journal*, 449, p. 137833.

Syed, Z., Sogani, M., Rajvanshi, J. and Sonu, K. 2023. "Microbial biofilms for environmental bioremediation of heavy metals: a review." *Applied Biochemistry and Biotechnology*, 195(9), pp. 5693–5711.

Tang, M., Liao, S., Qu, J., Liu, Y., Han, S., Cai, Z., Fan, Y., Yang, L., Li, S. and Li, L. 2022. "Evaluating bacterial pathogenesis using a model of human airway organoids infected with Pseudomonas aeruginosa biofilms." *Microbiology Spectrum*, 10(6), pp. e02408–e024022.

Tatsaporn, T. and Kornkanok, K. 2020. "Using potential lactic acid bacteria biofilms and their compounds to control biofilms of foodborne pathogens." *Biotechnology Reports*, 26, p. e00477.

Thallinger, B., Prasetyo, E.N., Nyanhongo, G.S. and Guebitz, G.M. 2013. "Antimicrobial enzymes: an emerging strategy to fight microbes and microbial biofilms." *Biotechnology Journal*, 8(1), pp. 97–109.

Tripathi, M., Singh, P., Singh, R., Bala, S., Pathak, N., Singh, S., Chauhan, R.S. and Singh, P.K. 2023. "Microbial biosorbent for remediation of dyes and heavy metals pollution: a green strategy for sustainable environment." *Frontiers in Microbiology*, 14, p. 1168954.

Ugya, Y.A., Chen, H., Sheng, Y., Ajibade, F.O. and Wang, Q. 2023. "A review of microalgae biofilm as an eco-friendly approach to bioplastics, promoting environmental sustainability." *Environmental Research*, 236, p. 116833.

Ünal Turhan, E., Erginkaya, Z., Korukluoğlu, M. and Konuray, G. 2019. "Beneficial biofilm applications in food and agricultural industry." In: Malik, A., Erginkaya, Z., and Erten, H. (eds.) *Health and Safety Aspects of Food Processing Technologies* (pp. 445–469). Springer, Cham, Springer Nature Switzerland.

Wang, Z.W. and Chen, S. 2009. "Potential of biofilm-based biofuel production." *Applied Microbiology and Biotechnology*, 83, pp. 1–18.

Willett, J.L., Ji, M.M. and Dunny, G.M. 2019. "Exploiting biofilm phenotypes for functional characterization of hypothetical genes in Enterococcus faecalis." *NPJ Biofilms and Microbiomes*, 5(1), p. 23.

Wu, T., Zhong, L., Ding, J., Pang, J.W., Sun, H.J., Ding, M.Q., Ren, N.Q. and Yang, S.S. 2023. "Microplastics perturb nitrogen removal, microbial community and metabolism mechanism in biofilm system." *Journal of Hazardous Materials*, 458, p. 131971.

Yin, W., Wang, Y., Liu, L. and He, J. 2019. "Biofilms: the microbial "protective clothing" in extreme environments." *International Journal of Molecular Sciences*, 20(14), p. 3423.

Zinicovscaia, I. and Balintova, M. 2023. "Environmentally-friendly materials in wastewater treatment." *Materials*, 16(18), p. 6181.

3 Microbial Strategies of Chrysene Remediation

Nitu Gupta, Apurba Koley, Sandipan Banerjee,
Binoy Kumar Show, Anudeb Ghosh, Sneha Banerjee,
Raza Rafiqul Hoque, and Srinivasan Balachandran

3.1 INTRODUCTION

Over the course of several decades, industrialization has steadily increased, releasing numerous toxic substances into our environment. Over the years, these substances have degraded the air, water, and soil quality and adversely affected our existence. PAHs are the most prominent omnipresent organic contaminants. PAHs are known to be immune suppressants, teratogens, carcinogens, and mutagens, among other serious health risks. Its hydrophobic properties allow it to assimilate quickly in the gastrointestinal tract of mammals (Barbosa et al., 2023; Gupta et al., 2024; Abdel-Shafy and Mansour, 2016). The cleanup in adding restoration strategies must, therefore, receive a significant priority to purge the damaged environment of these toxins. PAHs are viewed as one of the chief reasons for the ecological catastrophe. They are organic compounds represented by carbon and hydrogen atoms grouped to form 2–7 condensed or fused aromatic ring-like structures. They can be classified into the following two groups: low molecular weight PAHs (LMW-PAHs), which have fewer than four rings, and high molecular weight PAHs (HMW-PAHs), which consist of four or more rings (NCBI, 2023; Biswas and Ghosh, 2014).

Chrysene is HMW-PAHs comprised of four attached benzenoid rings, having a molecular weight of 228.29 g/mol and the chemical formula $C18H12$. It is categorized by the USEPA as a priority pollutant and is considered a human carcinogen (USEPA, 2002; Kaur et al., 2022). Chrysene is a crystalline, white substance composed of four fused benzene rings arranged symmetrically, resulting in incomplete combustion of organic matter. Similar to other PAHs, it undergoes chemical oxidation, accumulates in soils and aquifers, and can be volatilized, photolyzed, and adsorbed. Consequently, chrysene is widespread in various parts of the ecosystem, leading to numerous pathways of exposure to these carcinogens (Hussain et al., 2018). The intricate structural makeup and robust molecular arrangement that chrysene possesses make it extremely resistant under typical circumstances. Additionally, chrysene's low water solubility hinders its biodegradation and its persistence attributes in the nature (Kuppusamy et al., 2017) It has a variety of structural isomers, e.g., naphthalene, benzo(c)phenanthrene, benzo(a)anthracene, and triphenylene, depending on the position of the aromatic rings in their molecular structure. This structural variation causes carcinogenicity and toxicity attributes (Smith et al., 1989; Nwanna et al., 2006;

DOI: 10.1201/9781003407317-3

TABLE 3.1

Physical and Chemical Characteristics of Chrysene (Biswas and Ghosh, 2014)

Chemical Structure	Chemical Property	Description
Molecular structure of chrysene	Synonyms	• Chrysene • Benzo (a) phenanthrene • 1,2-Benzophenanthrene • 1,2,5,6-Dibenzonaphthalene
	Molecular formula	$C_{18}H_{12}$
	Molecular weight	228.3 g/mol
	Water solubility	0.002 mg/L
	Ethanol solubility	1 g/1300 mL
	Boiling point	448°C at 760 mmHg
	Melting point	255–256°C
	Vapor pressure	6.23×10^{-9} mmHg
	Log K_{ow}	5.9
	Log K_{oc}	6.11–7.34
	Density	1.274 g/cm³
	Physical state	Crystalline solid, particulate

Hadibarata et al., 2009; Dhote et al., 2010; Ghevariya et al., 2011; Nayak et al., 2011; Vaidya et al., 2018). Table 3.1 illustrates the physical and chemical characteristics of chrysene. Chrysene is a PAH, a class of highly cancer-causing substances. Due to the mutagenic properties of its metabolites, it is carcinogenic (Kaur et al., 2022). Chrysene is categorized as B2 by the EPA's Integrated Risk Management System (a potent human carcinogen) (Flowers et al., 2002). There are several approaches in use to eradicate PAHs from highly polluted environments, but bioremediation has emerged as the best choice. Its exceptional efficacy, versatility in addressing diverse contaminants, generation of environmentally harmless secondary byproducts, and utilization of sustainable green technology make it the unrivaled option (Biswas and Ghosh, 2014: Pramanik et al., 2018). Because of its refractory properties, such as high molecular weight, recalcitrance, carcinogenic aspects, sparse solubility etc., bioavailability and, therefore, the persistence of chrysene levels in the environment are very low, allowing it to prevent from microbial metabolism. When compared to physical and chemical techniques like soil washing, incineration chemical reduction and oxidation, vitrification, thermal desorption, immobilization, photocatalytic, and electrokinetic remediation, etc. Bioremediation consistently serves as a cheap, energy-efficient, and environmentally sustainable technique for cleanup (Mandree et al., 2021; Ismail et al., 2022).

3.2 BIBLIOGRAPHIC ANALYSIS

A comprehensive bibliographic review encompassed approximately 467 research papers imported from the Web of Science database, emphasizing the bioremediation

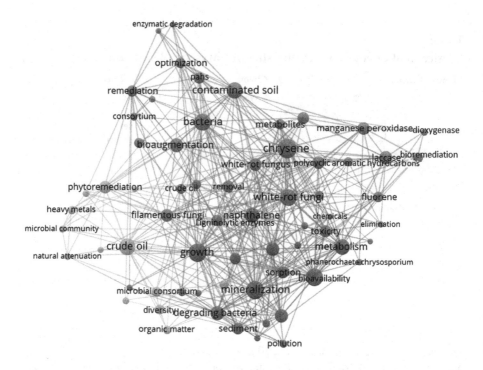

FIGURE 3.1 Bibliographical analysis of the co-occurrence of keywords using VOSviewer software.

of chrysene. Using VOSviewer software, all the co-occurrence keywords (seven co-occurrence keywords) have been screened and aligned in Linlog/modularity network, representing the most frequently searched terms in those research articles (Gupta et al., 2024; Koley et al., 2024; Koley et al., 2022), as depicted in Figure 3.1. This analysis of the co-occurrence of keywords discovered a total of four clusters. The anecdotal connections are connected by straight lines of a unique size and color that connect co-occurrence terms and all groups.

Cluster 1 (red cluster): microbial degradation can facilitate the mineralization of PAHs and diesel fuel in contaminated soil and sediments, leading to a reduction in pollution. Cluster 2 (green cluster): the enzymes, i.e., manganese peroxidase, laccase, and dioxygenase, found in fungi, are essential for the degradation of PAHs. Cluster 3 (blue cluster): through the optimization of the fungal and bacterial consortium, the biodegradation of chrysene as well as other PAHs in contaminated environment can be improved, leading toward the production of nontoxic metabolites. Clusters 4 (yellow cluster): PAHs, such as chrysene, can be found in sources like crude oil, partial combustion of organic matter, and oil spills, and its degradation can be achieved through the utilization of microbial communities and phytoremediation techniques. By conducting this study that utilizes text mining based on relevant keywords, a generalized map for the bioremediation of chrysene has been generated.

3.3 SOURCES OF CHRYSENE

Chrysene is frequently released into the environment due to incomplete combustion. Chrysene is believed to have been emitted into the environment, i.e., 93% in the atmosphere and the remaining 7% being equally distributed across land and water (Hussain et al., 2018). It has a strong correlation with sediments, soils, and particulate particles. Because of its high octanol–water partitioning coefficient value, chrysene is usually adsorbed into dust in the atmosphere. These molecules may be deposited by a variety of mechanisms or washed onto the soil by rain (NCBI, 2023). Chrysene, characterized by low water solubility and poor vapor pressure, tends to remain immobilized in soil without a significant movement. In aquatic environments, it is absorbed by particulate matter in riverbeds and sediments. Atmospheric deposition, resulting from emissions during fossil fuel combustion, industrial activities, and transportation, contributes to chrysene's presence in the soil through processes like dry and wet deposition. The dispersion of chrysene is influenced by the atmospheric mass movement and trans-border deposition, affecting both local and rural areas (Hussain et al., 2018).

The principal sinks for atmospheric chrysene are soil and road dust. It can be introduced into the nature primarily by both natural sources (volcanic eruptions and wildfires) and human activities (smoke from open fires, traditional cooking practices with wood, car exhaust, smoked, or grilled food, and cigarettes) (Yamini and Rajeswari, 2023; Yang et al., 2022; Li et al., 2022;). Petroleum products, such as petroleum, waxes, clarified oils, creosote, coal tar, etc., also contain significant amounts of chrysene. It is predominately exposed in the environment by pyrogenic processes, i.e., conversion of organic matter in an anaerobic condition, for example, thermal cracking of heavy hydrocarbons into small hydrocarbons and production of coal tar and coke by destructive distillation (Figure 3.2). The exposure to PAHs, typically through a mixture of chemically similar compounds, occurs via consumption of contaminated food or water, inhalation of polluted air in densely populated areas or car exhaust, and exposure through handling PAH-laden materials or employed in coal manufacturing unit (Ravanbakhsh et al., 2022).

3.4 BIOACTIVATION OF CHRYSENE

Before exhibiting the carcinogenic attributes, PAHs must undergo bioactivation metabolism, i.e., bay-site diol epoxides through dihydrodiols with a bay-site double bond via the synergistic effect of epoxide hydrolase and cytochrome P450-dependent monooxygenases (Conney, 1982). Chrysene undergoes bioactivation within the body through metabolic processes, leading to the formation of harmful and highly reactive intermediates that are capable of damaging DNA and other large molecules. Chrysene is biologically activated primarily via cytochrome (CYP) enzyme oxidation. The bay-region in chrysene aids the oxidases in breaking it down into reactive diol epoxides. Chrysene is also transformed by this mechanism into other reactive metabolites, such as other phenol and quinone molecules (Glatt et al., 1986). Chrysene can undergo activation through two pathways, resulting in the construction

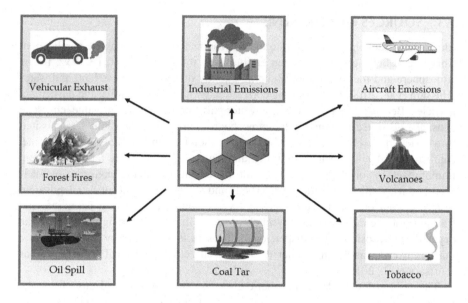

FIGURE 3.2 Major sources of chrysene.

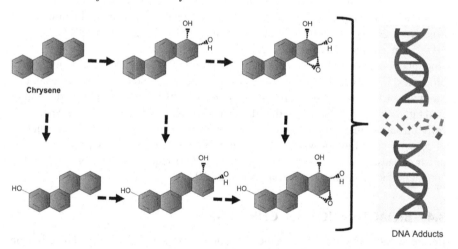

FIGURE 3.3 Chrysene activation via bay-region diol-epoxide and triol-epoxide pathways (Phillips and Grover, 1994).

of DNA-binding metabolites. One way involves the formation of chrysene bay-region diol-epoxide, while the other pathway leads to the production of triol-epoxides, which contain a phenolic hydroxyl group as illustrated in Figure 3.3. Triol-epoxides, including 9-hydroxychrysene-1,2-diol 3,4-oxide, have been synthesized and interacted with DNA, demonstrating mutagenic characteristics. The exact in vivo processes leading to the formation of triol-epoxide-nucleic acid adducts are not yet completely understood (Phillips and Grover, 1994)

The mutagenic and carcinogenic properties of chrysene are linked to its metabolites, such as 1,2-dihydrodiol and 1,2-diol-3,4-epoxide. The reactive metabolites that are produced can subsequently go through other processes, such as epoxidation, the creation of dihydrodiols, and conjugation with glutathione and hydroxyalkyl, quinones, and other cellular components, which are well-known for forming DNA adducts and inducing genotoxicity (Luch and Baird, 2005; Moubarz et al., 2022). According to the metabolic research, chrysene undergoes microsomal oxidation to produce monohydroxy derivatives such as 1- and 3-phenols, as well as 1,2-, 3,4-, and 5,6-dihydrodiols (International Agency for Research on Cancer (IARC), 1983). Chrysene trans-3,4-dihydrodiol is the primary metabolite formed during the subsequent metabolism of these dihydrodiol intermediates, which also leads to the generation of 1,2-diol-3,4-epoxide and 3,4-diol-1,2-epoxide (Vondracek and Machala, 2021).

PAHs are recognized to attach to and trigger the aryl hydrocarbon receptor transcription factor. Binding of the receptor to xenobiotic response elements leads to the activation of various xenobiotic metabolism-related enzymes. These include phase II enzymes such as glutathione-S-transferase (GST) A2, UDP-glucuronosyltransferase (UGT) 1A1, and NADPH:quinone oxidoreductase (NQO1), as well as cytochrome P450 (CYP) enzymes like CYP1A1, CYP1A2, and CYP1B1 (Tao et al., 2021). In general, the bioactivation of chrysene involves its oxidation by CYP enzymes, resulting in the creation of active metabolites that are able to damage DNA and other biomolecules. The formation of dihydrodiol metabolites and their subsequent conversion to diol epoxides is a critical step in this process. Additional research is required to comprehensively elucidate the mechanisms of chrysene bioactivation and the variables affecting its metabolism.

3.5   CHRYSENE BIOREMEDIATION

The industrialization boom overloaded the environment with PAHs such as chrysene compounds. To restore the ecosystem to nontoxic soil conditions, it is necessary to reclaim the PAH-contaminated areas through the breakdown of PAH molecules into nontoxic end bioproducts, such as water, inorganic minerals, and carbon dioxide (Banerjee et al., 2016; Thacharodi et al., 2023). Different methods have been employed to remediate PAHs including immobilization methods (such as capping, stabilization, dredging, and excavation), mobilization techniques (such as thermal desorption, washing, electrokinetics, and surfactant-assisted methods), and biological degradation techniques (Kumar et al., 2021). However, bioremediation emerges as the preferred and most effective choice among these methods. Throughout the past few decades, there has been much documentation of the PAH mineralization (Yamini and Rajeswari, 2023; Thacharodi et al., 2023; Banerjee et al., 2022; Cerniglia, 1997). However, there is limited literature as far as the concern with individual PAH bioremediation, such as chrysene. Microorganisms like bacteria and fungi are getting greater attention among the numerous forms of chrysene bioremediation approaches due to their vast availability, cheapness, and environmental friendliness.

3.5.1 BACTERIAL-MEDIATED CHRYSENE REMEDIATION

Bacteria are pivotal in breaking down organic pollutants in polluted environments. Various bacterial species are known for their ability to degrade PAHs, with many effective strains being isolated from polluted soils or sediments. *Mycobacterium* spp., *Pseudomonas aeruginosa*, *P. fluorescens*, *Haemophilus* spp., *Paenibacillus* spp., and *Rhodococcus* spp. are some of the frequently studied bacteria known for degrading PAHs (Premnath et al., 2021). The breakdown of PAHs is influenced by factors such as the chemical molecule's nature and molecular structure, the types and quantities of microorganisms present, and the persistent environmental conditions. They are transformed into water, carbon dioxide, or methane under anaerobic conditions, and they undergo biodegradation or biotransformation into simpler metabolites (Haritash and Kaushik, 2009). Microorganisms can cometabolize contaminants or use them as substrates depending on a variety of factors (Rafeeq et al., 2023). Consequently, defining key elements for effective pollution remediation requires an understanding of catabolic pathways, processes, and relevant enzymes. (Alexander, 1999; Seo et al., 2009). A comprehensive list of chrysene-degrading bacteria is provided, highlighting their specific enzymatic mechanisms and the percentage of chrysene degradation achieved by each bacterium as shown in Table 3.2.

Chrysene undergoes oxidation where an oxygen molecule is added to a benzenoid ring, facilitated by the dioxygenase activity, forming a cis-dihydrodiol intermediate. This intermediate subsequently proceeds to breakdown leading to complete degradation (Dhote et al., 2010). Dioxygenases are crucial enzymes in the breakdown of PAHs. The existence of catabolism-related genes that encodes these enzymes in microorganisms indicates their role in biodegradation processes. Thomas et al. (2021) conducted research on chrysene, where the compound was found to degrade 90% by *Bacillus halotolerans*, with the involvement of enzymes, such as laccase and catechol 1,2 dioxygenase. *Bacillus* sp. *and Pseudomonas* sp. degraded 15% and 17% of the compound, respectively, when catechol 1,2-dioxygenase and catechol 2,3-dioxygenase acted upon it (Dhote et al., 2010). Ghevariya et al. (2011) found that 83.96% of chrysene could be degraded when *Achromobacter xylosoxidans* was involved. Pyrene and chrysene may be used as the only sources of carbon by the *Rhodococcus* sp. strain UW1 that was isolated from polluted soil (Kanaly and Harayama, 2000). Willison (2004) reported the potentiality of Sphingomonas to detoxify four-ring PAH, such as chrysene, with a degradation rate of 97.5% after 35 days. Nayak et al. (2011) proposed the chrysene biodegradation pathway by *Pseudoxanthomonas* sp. PNK-04. The suggested intermediates were hydroxyphenyl anthroic acid catabolized into 1-hydroxy-2-naphthoic acid, further broken into salicylic acid, and then catechol, which ultimately enters into the tricarboxylic acid cycle as illustrated in Figure 3.4.

3.5.2 FUNGAL-MEDIATED CHRYSENE REMEDIATION

Compared to bacteria, fungi are more plastic, more tolerant, totipotent (regenerate from hyphal fragments as well as spores), and via hyphal elongation, they are able to penetrate or reach xenobiotics (Cerniglia and Sutherland, 2001). The fungi that

TABLE 3.2
Bacterial Strategies for Bioremediation of Chrysene

Compounds	Bacteria	Phylum	Mechanism	Degradation (%)	Reference
Chrysene	*Bacillus* sp.	Firmicutes	Phytoremediation (*Pennisetum purpureum*)	97.54	Riskuwa-Shehu et al. (2022)
	Micrococcus sp.	Actinobacteria			
	Pseudomonas aeruginosa	Proteobacteria			
	Klebsiella pneumoniae	Proteobacteria			
	Flavobacterium sp.	Bacteroidetes			
Chrysene	*Pseudomonas aeruginosa*	Proteobacteria	Biosurfactant-assisted biodegradation	63	Chirwa et al. (2021)
	Acinetobacter sp.	Proteobacteria			
Chrysene	*Streptomyces* sp.	Actinobacteria	*In situ* application of microbiological consortium	59.9	Roszak et al. (2021)
	Arthrobacter sp.	Proteobacteria			
	Pseudomonas sp.	Actinobacteria			
	Achromobacter sp.	Proteobacteria			
	Microbacterium sp.	Proteobacteria			
	Capriavidus sp.	Actinobacteria			
	Rhodococcus sp.	Actinobacteria			
Chrysene	*Bacillus halotolerans*	Firmicutes	Enzymatic activity (laccase and catechol 1,2 dioxygenase)	90	Thomas et al. (2021)
Chrysene	*Pseudomonas putida*	Proteobacteria	Meta-cleavage pathway	60	Ahmad et al. (2020)
	Pseudoxanthomonas sp.	Proteobacteria			
Chrysene	*Parvibaculum* sp.	Proteobacteria	Microbial-assisted phytoremediation (*Dactylis glomerate*)	38	Borowik et al. (2020)
	Rhodococcus sp.	Actinobacteria			
Chrysene	*Epibacterium* sp.	Proteobacteria	Enzymatic activity	20	Dell'Anno et al. (2020)

(Continued)

TABLE 3.2 (Continued)
Bacterial Strategies for Bioremediation of Chrysene

Compounds	Bacteria	Phylum	Mechanism	Degradation (%)	Reference
Chrysene	P. aeruginosa	Proteobacteria		72	Olowomofe et al. (2019)
	Bacillus cereus	Firmicutes	Enzymatic activity	75	
	Dyadobacter koreensis	Bacteroidetes		85	
Chrysene	Rhodococcus sp.	Actinobacteria	Synergistic degradation with bioaugmentation	96	Vaidya et al. (2018)
	Bacillus sp.	Firmicutes			
	Burkholderia sp.	Proteobacteria			
Chrysene	Pseudomonas sp.	Proteobacteria	Synergistic degradation	66.45	Dave et al. (2014)
	Achromobacter xylosoxidans	Proteobacteria			
	Sphingomonas sp.	Proteobacteria			
Chrysene	Pseudomonas sp.	Proteobacteria	Enzymatic activity	35–69	Nwinyi et al. (2013)
Chrysene	Achromobacter xylosoxidans	Proteobacteria	Enzymatic activity (Lignin modifying enzymes)	85.96	Ghevariya et al. (2011
Chrysene	Pseudoxanthomonas sp.	Proteobacteria	Enzymatic activity (Hydroxylase, dioxygenase, and dehydrogenase)	60	Nayak et al. (2010)
Chrysene	Novosphingobium pentaromativorans	Proteobacteria	Enzymatic activity	80	Lafortune et al. (2009)
Chrysene	Mycobacterium parmense	Proteobacteria	Enzymatic activity	53	Lladó et al. (2009)
	Pseudomonas mexicana	Proteobacteria			
Chrysene	Bacillus sp.	Firmicutes	Enzymatic activity	15.0	Dhote et al. (2010)
	Pseudomonas sp.	Proteobacteria	(Dioxygenase)	17.0	
Chrysene	Sphingomonas sp.	Proteobacteria	Biofilm activity	97.5	Willison (2004)

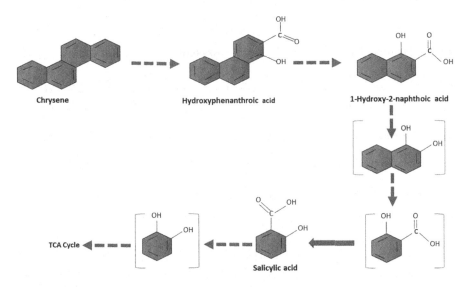

FIGURE 3.4 Chrysene biodegradation pathway by *Pseudoxanthomonas* sp. PNK-04 (Nayak et al., 2011).

have proven to be effective in breaking down chrysene include *Aspergillus niger, Trametes versicolor, and Pleurotus ostreatus,* among others, which act through different mechanisms such as oxidation, reduction, and enzymatic activity (Gupta et al., 2024; Aydin et al., 2017; Pinedo-Rivilla et al., 2009; Cerniglia and Sutherland, 2001;). There are two types of fungi, namely, ligninolytic and non-ligninolytic. A significant fungus group known as non-ligninolytic uses the enzymes, such as cytochrome P450 monooxygenase, dioxygenase, and epoxide hydrolase, to degrade PAHs (Kumari et al., 2021). The ligninolytic fungi secrete ligninolytic enzymes that breakdown PAHs through nonspecific oxidative radicals, such as laccase and peroxidase enzymes (a group of organisms, primarily basidiomycetes, that releases enzymes that transform the lignin in wood). Although non-ligninolytic enzymes can also be produced by ligninolytic fungi, it is not yet known which specific enzymes help in breakdown of PAHs (Raghukumar et al., 2006). Hidayat et al. (2012) documented that *Fusarium* sp. is endowed with the potential to degrade chrysene by approximately 48% degradation rate in 30 days. Ascomycota, primarily non-ligninolytic fungi, are widely recognized as a promising bioagent for PAH remediation. They are followed by Basidiomycota, which are ligninolytic fungi, and finally by Zygomycota, classified among non-ligninolytic fungal groups (Passarini et al., 2011; Banerjee and Mandal, 2020). Table 3.3 presents a summary of the ability of fungi to biodegrade chrysene, detailing their phyla, biodegradation mechanisms, and the percentage of removal. Another such research by Al Farraj et al. (2019) found the results to be 77% of chrysene that was degraded when *Hortaea* sp. B15 acted upon the compound.

Hadibarata et al. (2009) reported that the *Polyporus* sp. S133 is able to mineralize chrysene with a degradation rate of 65%, incubated with 10% polypeptone at 120 rpm for 30 days. Through the identification of multiple metabolites such as chrysenequinone, 1-hydroxy-2-naphthoic acid, phthalic acid, salicylic acid, protocatechuic acid,

TABLE 3.3
Fungal Strategies for Bioremediation of Chrysene

Fungus	Phylum	Mechanism	Degradation Percentage (%)	Reference
Trametes versicolor	Basidiomycota	Laccase	75.80	Vipotnik et al. (2022)
Agrocybe aegerita	Basidiomycota	Ligninolytic enzymes	89	Vipotnik et al. (2021)
Hortaea sp.	Ascomycota	Enzymatic activity (dioxygenase)	77	Al Farraj et al. (2019)
Pleurotus sajor-caju	Basidiomycota	Ligninolytic enzymes	NA	Saiu et al. (2016)
Fusarium sp.	Ascomycota	Ligninolytic enzymes	48	Hidayat and Tachibana (2015)
Fusarium sp.	Ascomycota	1,2-dioxygenase, 2,3-dioxygenase	35	Hidayat et al. (2012)
Cochliobolus lunatus	Ascomycota	Laccase, manganese peroxidase, and cytochrome P450	93.10	Bhatt et al. (2014)
Polyporus sp.	Basidiomycota	Dioxygenase, laccase, lignin, and manganese peroxidase	65	Hadibrata et al. (2009)
Trichocladium canadense	Ascomycota	Ligninolytic activity	40.90	Silva et al. (2009)
Pichia anomala	Ascomycota	Cometabolism	75.90	Hesham et al. (2006)
Cladosporium sphaerospermum	Ascomycota	Laccase activity	NA	Potin et al. (2004)
Penicillum sp.	Ascomycota	Laccase activity	30	Zheng and Obbard (2003)

NA, not available.

gentisic acid, and catechol, researchers elucidated the chrysene breakdown process by *Polyporus* sp. S133. Also, the presence of various enzymes in *Polyporus* sp. S133, such as 1,2-dioxygenase, 2,3-dioxygenase, manganese peroxidase (MnP), lignin peroxidase (LiP), and laccase is detected. Such enzymes use substrates like inorganic metal or organic complexes and molecular oxygen as electron acceptors (Zimmerman et al., 2008). Laccases also use oxygen as an oxidizing agent and convert it into water molecules (Tavares et al., 2006). *Trametes versicolor*, which was isolated from the soil and produced 37.8 U/g of laccase, accomplished 81.0% degradation of chrysene after a period of 3 weeks (Vipotnik et al., 2022). According to how they interact with their substrates, the peroxidases are divided into two groups as follows: MnP, which is the most efficient reducing substrate, and LiP, which catalyzes both aliphatic and aromatic substances. During metabolic processes, fungal producers synthesize

MnP and LiP, which require hydrogen peroxide (H_2O_2) for activation. H_2O_2 serves as an oxidizing agent that transforms PAHs into a substrate by oxidizing them. It is a crucial component of the fungal metabolic process that enables the degradation of PAHs. The peroxidase isozyme family includes versatile peroxidase (VPs), which possess the capability to operate on a diverse array of substrates. These enzymes are capable of catalyzing reactions on various substances, similar to other members of the peroxidase isozyme family (Deshmukh et al., 2016). Vipotnik et al. (2021) reported that *Agrocybe aegerita* accomplished 78% degradation of chrysene after 2 weeks of incubation, with laccase activity levels measuring 7.43 U/g. Furthermore, after 4 weeks of incubation (pH 7), MnP activity levels reached 7.21 U/g, and after 5 weeks of fermentation at pH 5, LiP activity levels were 2.24 U/g. The white rot fungi (WRF) utilize cytochrome P450 monooxygenase systems to eliminate numerous xenobiotic compounds, particularly organic pollutants, such as PAHs, through oxidative decontamination. This enzyme converts water-insoluble substances into water-soluble substances by incorporating molecular oxygen, which facilitates the breakdown of the organic pollutants. Many non-ligninolytic fungi can secrete this type of enzyme. Another study found that WRF *P. ostreatus* can degrade a substantial amount of PAHs, including 80% of all PAHs, 29.7% of chrysene, 69.1% of benzo(a)anthracene, 32.8% of benzo(k)fluoranthene, and 39.7% of benzo(b)fluoranthene (Covino et al., 2010). By providing valuable insights into the efficiency of fungal-mediated chrysene biodegradation, Table 3.2 shows a comprehensive list of chrysene-degrading fungi, along with their specific enzymatic mechanisms and percentage of chrysene degradation achieved.

3.6 FACTORS AFFECTING THE CHRYSENE BIOREMEDIATION

The biodegradation of xenobiotics, including its rate, is affected by environmental conditions, compound types, quantities of microbes, and the chemical nature and composition of the substance being degraded. Therefore, when designing a bioremediation system, several factors must be considered. Significant research has concentrated on exploring the capacity of microbes to mineralize xenobiotics such as PAHs. Numerous variables, including temperature, pH, microbial population, oxygen availability, acclimation capacity, nutrient availability, molecular assembly of the compound, and cellular transport features, impact the extent and speed of biodegradation (Singh and Ward, 2004; Haritash and Kaushik, 2009). Optimal temperature, ideal pH, bacterial strain, and the specific PAH molecule influence the rate of PAH breakdown.. Bacteria require certain nutrients to grow and metabolize PAHs. Furthermore, the addition of nutrients or other substances into the polluted area can promote the proliferation of PAHs degrading microbes and increase the efficiency of PAH biodegradation, i.e., bioaugmentation. Al-Dossary et al. (2021) identified several factors that can affect PAH degradation, including optimal pH and temperature for *Planomicrobium alkanoclasticum* that were observed to be 7.5 at 35°C supplemented with nutrients like $NaNO_3$ for better biodegradation, i.e., 91.4% degradation of 16 PAHs.

Temperature and pH are critical factors influencing biochemical and enzymatic processes. In a study by Vaidya et al. (2018) focusing on chrysene degradation, a consortium composed of *Rhodococcus* sp., *Bacillus* sp., and *Burkholderia* sp. demonstrated that optimal degradation at neutral pH, with efficient degradation, was also observed under acidic conditions. Specifically, at pH 6, the consortium degraded 64% of chrysene within 7 days. The degradation efficiency increased to 90% at pH 7.0 but declined rapidly as pH levels became more alkaline, with degradation efficiencies dropping approximately 3.6-fold at pH 9.0 and about 4.5-fold at pH 10. The consortium's activity was also found to be temperature-dependent, with optimal degradation occurring between 30°C and 37°C. Previous studies have similarly reported enhanced degradation of various PAHs under near-neutral pH conditions (Leahy and Colwell, 1990; Hambrick et al., 1980; Dibble and Bartha, 1979; Verstraete et al., 1976). Furthermore, degradation was observed to be most efficient between 30°C and 37°C, justifying previous studies, which conclude that, as for most aerobic enzymes, the optimum reaction temperature lies in the mentioned range. Often, the key enzyme-related activities for different metabolic reactions necessary for the growth of microorganisms are decreased or inhibited by the presence of heavy metals (Chen et al., 2008; Hamzah et al., 2011; Mukherjee and Roy, 2013; Patel et al., 2012a, 2012b; Vaidya et al., 2018). For bacteria to break down a particular PAH, it needs to be made accessible for absorption. PAHs become available for biological processes when they are either dissolved in a liquid or present in the gaseous phase. When PAHs adhere to soil grains, they become less accessible to bacteria because they are separated from the enzymes that bacteria use to degrade them. Therefore, the bioavailability of a particular PAH is influenced by its solubility, which is a critical determinant in its accessibility for microbial degradation. The molecular weights of PAHs have a significant impact on how soluble they are in the water. Competitive inhibition may potentially have an effect on PAH decomposition. Conversely, if bacteria involved in PAH degradation encounter a molecule that serves as a more readily available food source, the rate of PAH decomposition may decrease. When the active sites of enzymes that bacteria employ to break down PAHs as a carbon source are nonspecific, competitive inhibition happens. These general enzymes can bind to a wide range of various substances (Abdel-Shafy and Mansour, 2016). A thorough comprehension of the factors influencing the biodegradation of chrysene is crucial for designing effective bioremediation strategies for contaminated sites. By managing these factors, such as adjusting environmental factors to suit the microbial community and providing the required nutrients, the efficiency of chrysene biodegradation can be significantly improved. This method can significantly contribute to restoring contaminated environments and mitigating the potential risks associated with chrysene exposure, thereby benefiting both environmental quality and public health. The investigation of genetic engineering and other cutting-edge methods to improve biodegradation efficiency may fill a future gap. This could involve manipulating specific genes or enzymes involved in chrysene degradation pathways or engineering novel microbial strains with improved biodegradation capabilities using synthetic biology. Additionally, investigating the potential of bioremediation techniques such as phytoremediation or microbial fuel cells could also offer promising avenues for future research (Nzila and Musa, 2021).

3.7 CONCLUSIONS

Chrysene, classified as a PAH, enters the atmosphere due to various human activities. It is a hazardous and persistent contaminant that seriously endangers both the environment and human health. Chrysene can be effectively removed from contaminated areas by biodegradation. Research has revealed that diverse microorganisms, encompassing bacteria, fungi, and algae, mineralize chrysene using various metabolic pathways. Chrysene is largely biodegraded by an oxidative degradation process, where it first undergoes oxidation to form chrysenequinone, which is subsequently transformed into less harmful intermediates, including phthalic acid and then 2-hydroxy-1-naphthoic acid. The biodegradation of chrysene can be influenced by various parameters, including temperature, pH, nutrient availability, and the occurrence of other contaminants. The rate of chrysene mineralization can be enhanced by incorporating bioaugmentation and biostimulation techniques. Chrysene toxicity is influenced by the exposure period, its doses, and organism sensitivity. It is designated as a potential human carcinogen and associated with various health issues, e.g., cancer, skin irritation, and respiratory issues. Through biodegradation, microbes can efficiently transform chrysene into less hazardous chemicals, hence lowering toxicity.

ACKNOWLEDGMENTS

Nitu Gupta acknowledged the financial support provided under the JRF scheme by the University Grants Commission. Sandipan Banerjee expressed gratitude to the Department of Biotechnology, Government of India, for the financial support through the DBT Twinning Project and Research Fellowship [No. BT/PR25738/ NER/95/1329/2017 dated December 24, 2018].

REFERENCES

Abdel-Shafy, H. I. and M. S. Mansour. 2016. A review on polycyclic aromatic hydrocarbons: source, environmental impact, effect on human health and remediation. *Egypt. J. Pet.* 25(1): 107–123. https://doi.org/10.1016/j.ejpe.2015.03.011

Ahmad, F., D. Zhu, and J. Sun. 2020. Bacterial chemotaxis: a way forward to aromatic compounds biodegradation. *Environ. Sci. Eur.* 32: 1–18. https://doi.org/10.1186/s12302-020-00329-2

Al Farraj, D. A., T. Hadibarata, A. Yuniarto, A. Syafiuddin, H. K. Surtikanti, M. S. Elshikh, and R. Al-Kufaidy. 2019. Characterisation of pyrene and chrysene degradation by halophilic Hortaea sp. B15. *Bioprocess. Biosyst. Eng.* 42: 963–969. https://doi.org/10.1007/s00449-019-02096-8

Al-Dossary, M. A., S. A. Abood, and H. T. Al-Saad. 2021. Effects of physicochemical factors on PAH degradation by *Planomicrobium alkanoclasticum*. *Remediation J.* 31(2): 29–37. https://doi.org/10.1002/rem.21673

Alexander, M. 1999. *Biodegradation and bioremediation*. San Diego: Elsevier Science.

Aydin, S., H. A. Karaçay, A. Shahi, S. Gökçe, B. Ince, and O. Ince. 2017. Aerobic and anaerobic fungal metabolism and Omics insights for increasing polycyclic aromatic hydrocarbons biodegradation. *Fungal Biol. Rev.* 31(2): 61–72. https://doi.org/10.1016/j.fbr.2016.12.001

Banerjee, S., and N. C. Mandal. 2020. Fungal Bioagents in the remediation of degraded soils. In: Singh, J. S., & Vimal, S.R., (eds.), *Microbial services in restoration ecology* (pp. 191–205). Amsterdam: Elsevier. https://doi.org/10.1016/B978-0-12-819978-7.00013-0

Banerjee, S., T. K. Maiti, and R. N. Roy. 2016. Identification and product optimization of amylolytic *Rhodococcus opacus* GAA 31.1 isolated from gut of *Gryllotalpa africana*. *J. Genet. Eng. Biotechnol.* 14(1): 133–141. https://doi.org/10.1016%2Fj.jgeb.2016.05.005

Banerjee, S., T. K. Maiti, and R. N. Roy. 2022. Enzyme producing insect gut microbes: an unexplored biotechnological aspect. *Crit. Rev. Biotechnol.* 42(3): 384–402. https://doi.org/10.1080/07388551.2021.1942777

Barbosa Jr, F., B. A. Rocha, M. C. Souza, M. Z. Bocato, L. F. Azevedo, J. A. Adeyemi,... A. D. Campiglia. (2023). Polycyclic aromatic hydrocarbons (PAHs): updated aspects of their determination, kinetics in the human body, and toxicity. *J. Toxicol. Environ. Health - B: Crit. Rev.* 1–38. https://doi.org/10.1080/10937404.2022.2164390

Bhatt, J. K., C. M. Ghevariya, D. R. Dudhagara, R. K. Rajpara, and B. P. Dave. 2014. Application of response surface methodology for rapid chrysene biodegradation by newly isolated marine-derived fungus Cochliobolus lunatus strain CHR4D. *J. Microbiol.* 52: 908–917. https://doi.org/10.1007/s12275-014-4137-6

Biswas, S. and B. Ghosh. 2014. *Chrysene. Encyclopedia of toxicology*, Wexler, P. (ed.) (third edition, pp. 959–962). Bethesda, MD: Academic Press. https://doi.org/10.1016/B978-0-12-386454-3.00286-4

Borowik, A., J. Wyszkowska, M. Kucharski, and J. Kucharski. 2020. The role of Dactylis glomerata and diesel oil in the formation of microbiome and soil enzyme activity. *Sensors.* 20(12): 3362. https://doi.org/10.3390/s20123362

Cerniglia, C. E. 1997. Fungal metabolism of polycyclic aromatic hydrocarbons: past, present and future applications in bioremediation. *J. Ind. Microbiol.* 19(5–6): 324–333. https://doi.org/10.1038/sj.jim.2900459

Cerniglia, C. E., and J. B. Sutherland. 2001. Bioremediation of polycyclic aromatic hydrocarbons by ligninolytic and non-ligninolytic fungi. In: G. M. Gadd (ed.), *British mycological Society Symposium Series* (Vol. 23, pp. 136–187). Cambridge: Cambridge University Press. https://doi.org/10.1017/CBO9780511541780

Chen, Y., J. J. Cheng, and K. S. Creamer. 2008. Inhibition of anaerobic digestion process: a review. *Bioresour. Technol.* 99(10): 4044–4064. https://doi.org/10.1016/j.biortech.2007.01.057

Chirwa, E. M. N., T. B. Lutsinge-Nembudani, O. M. Fayemiwo, and F. A. Bezza. 2021. Biosurfactant assisted degradation of high molecular weight polycyclic aromatic hydrocarbons by mixed cultures from a car service oil dump from Pretoria Central Business District (South Africa). *J. Clean. Prod.* 290: 125183.https://doi.org/10.1016/j.jclepro.2020.125183

Conney, A. H. 1982. Induction of microsomal enzymes by foreign chemicals and carcinogenesis by polycyclic aromatic hydrocarbons: GHA Clowes Memorial Lecture. *Cancer Res.* 42(12): 4875–4917.

Covino, S., K. Svobodová, M. Čvančarová, A. D'Annibale, M. Petruccioli, F. Federici,... T. Cajthaml. 2010. Inoculum carrier and contaminant bioavailability affect fungal degradation performances of PAH-contaminated solid matrices from a wood preservation plant. *Chemosphere.* 79(8): 855–864. https://doi.org/10.1016/j.chemosphere.2010.02.038

Dave, B. P., C. M. Ghevariya, J. K. Bhatt, D. R. Dudhagara, and R. K. Rajpara. 2014. Enhanced biodegradation of total polycyclic aromatic hydrocarbons (TPAHs) by marine halotolerant Achromobacter xylosoxidans using Triton X-100 and β-cyclodextrin – a microcosm approach. *Mar. Pollut. Bull.* 79(1–2): 123–129. https://doi.org/10.1016/j.marpolbul.2013.12.027

Dell'Anno, F., C. L. J. Brunet, van Zyl, M. Trindade, P. N. Golyshin, A. Dell'Anno,... C. Sansone. 2020. Degradation of hydrocarbons and heavy metal reduction by marine bacteria in highly contaminated sediments. *Microorganisms.* 8(9): 1402. https://doi. org/10.3390/microorganisms8091402

Deshmukh, R., A. A. Khardenavis, and H. J. Purohit. 2016. Diverse metabolic capacities of fungi for bioremediation. *Indian J. Microbiol.* 56: 247–264. https://doi.org/10.1007/ s12088-016-0584-6

Dhote, M., A. Juwarkar, A. Kumar, G. S. Kanade, and T. Chakrabarti. 2010. Biodegradation of chrysene by the bacterial strains isolated from oily sludge. *World J. Microbiol.* 26: 329–335. https://doi.org/10.1007/s11274-009-0180-6

Dibble, J. T., and R. Bartha. 1979. Effect of environmental parameters on the biodegradation of oil sludge. *Appl. Environ. Microbiol.* 37(4): 729–739.

Flowers, L., S. H. Rieth, V. J. Cogliano, G. L. Foureman, R. Hertzberg, E. L. Hofmann, ... R. S. Schoeny. 2002. Health assessment of polycyclic aromatic hydrocarbon mixtures: current practices and future directions. *Polycyclic Aromatic Compounds.* 22(3–4): 811–821.

Ghevariya, C. M., J. K. Bhatt, and B. P. Dave. 2011. Enhanced chrysene degradation by halo-tolerant Achromobacter xylosoxidans using response surface methodology. *Bioresour. Technol.* 102(20): 9668–9674. https://doi.org/10.1016/j.biortech.2011.07.069

Glatt, H., A. Seidel, W. Bochnitschek, H. Marquardt, H. Marquardt, R. M. Hodgson,... F. Oesch, 1986. Mutagenic and cell-transforming activities of triol-epoxides as compared to other chrysene metabolites. *Cancer Res.* 46(9): 4556–4565.

Gupta, N., S. Banerjee, A. Koley, A. Basu, N. Gogoi, R. R. Hoque, N. C. Mandal, and S. Balachandran. 2024. Fungal strategies for the remediation of polycyclic aromatic hydrocarbons. In: Malik A., & Garg V.K. (eds.), *Bioremediation for sustainable environmental cleanup* (pp. 86–108). Boca Raton, FL: CRC Press.

Hadibarata, T., S. Tachibana, and K. Itoh. 2009. Biodegradation of chrysene, an aromatic hydrocarbon by Polyporus sp. S133 in liquid medium. *J. Hazard. Mater.* 164(2–3): 911–917. https://doi.org/10.1016/j.jhazmat.2008.08.081

Hambrick III, G. A., R. D. DeLaune, and W. H. Patrick Jr. 1980. Effect of estuarine sediment pH and oxidation-reduction potential on microbial hydrocarbon degradation. *Appl. Environ. Microbiol.* 40(2): 365–369.

Hamzah, A., A. Tavakoli, and A. Rabu. 2011. Detection of toluene degradation in bacteria isolated from oil contaminated soils. *Sains Malays.* 40(11): 1231–1235.

Haritash, A. K., and C. P. Kaushik. 2009. Biodegradation aspects of polycyclic aromatic hydrocarbons (PAHs): a review. *J. Hazard. Mater.* 169(1–3): 1–15. https://doi.org/10.1016/j. jhazmat.2009.03.137

Hesham, A. E. L., Z. Wang, Y. Zhang, J. Zhang, W. Lv, and M. Yang. 2006. Isolation and identification of a yeast strain capable of degrading four and five ring aromatic hydrocarbons. *Ann. Microbiol.* 56: 109–112. https://doi.org/10.1007/BF03174990

Hidayat, A., and S. Tachibana. 2015. Simple screening for potential chrysene degrading fungi. *KnE Life Sci.* 364–370. https://doi.org/10.18502/kls.v2i1.177

Hidayat, A., S. Tachibana, and K. Itoh. 2012. Determination of chrysene degradation under saline conditions by Fusarium sp. F092, a fungus screened from nature. *Fungal Biol.* 116(6): 706–714.

Hussain, K., R. R. Hoque, S. Balachandran, S. Medhi, M. G. Idris, M. Rahman, and F. L. Hussain. 2018. Monitoring and risk analysis of PAHs in the environment. In: Hussain C.M. (ed.), *Handbook of environmental materials management* (pp. 1–35). New Jersey: Springer Cham.

International Agency for Research on Cancer (IARC). 1983. Polynuclear aromatic compounds, part I, chemical, environmental and experimental data: chrysene. In: *IARC monographs on the evaluation of the carcinogenic risk of chemicals to humans* (Vol. 32, pp. 247–262). Lyon, France: International Agency for Research on Cancer. Available from: https://publications.iarc.fr/50 (accessed 02 March, 2022).

Ismail, N. A., N. Kasmuri, and N. Hamzah. 2022. Microbial bioremediation techniques for polycyclic aromatic hydrocarbon (PAHs) *Water Air Soil Pollut.* 233(4): 124. https://doi.org/10.1007/s11270-022-05598-6

Kanaly, R. A., and S. Harayama. 2000. Biodegradation of high-molecular-weight polycyclic aromatic hydrocarbons by bacteria. *J. Bacteriol. Res.* 182(8): 2059–2067.

Kaur, S., B. Kumar, P. Chakraborty, V. Kumar, and N. C. Kothiyal. 2022. Polycyclic aromatic hydrocarbons in PM10 of a north-western city, India: distribution, sources, toxicity and health risk assessment. *Int. J. Environ. Sci. Technol.* 1–16. https://doi.org/10.1007/s13762-021-03450-8

Koley, A., D. Bray, S. Banerjee, S. Sarhar, R. G. Thahur, A. K. Hazra, ... S. Balachandran. 2022. Water hyacinth (Eichhornia crassipes) a sustainable strategy for heavy metals removal from contaminated waterbodies. In: Malik, A., Kidwai, M.K., & Garg, V.K. (eds.), *Bioremediation of toxic metal(loid)s* (pp. 95–114). Boca Raton, FL: CRC Press.

Koley, A., A. Ghosh, S. Banerjee, N. Gupta, R. G. Thakur, B. K. Show, ... S. Balachandran. 2024. Phytoremediation of wastewater discharged from paper and pulp, textile and dairy industries using water hyacinth (Eichhornia crassipes). In: Malik, A., & Garg, V.K. (eds.) *Bioremediation for sustainable environmental cleanup* (pp. 238–261). Boca Raton, FL: CRC Press.

Kumar, M., N. S. Bolan, S. A. Hoang, A. D. Sawarkar, T. Jasemizad, B. Gao, ... J. Rinklebe. 2021. Remediation of soils and sediments polluted with polycyclic aromatic hydrocarbons: to immobilise, mobilise, or degrade? *J Hazard Mat.* 420: 126534. https://doi.org/10.1016/j.jhazmat.2021.126534

Kumari, R., A. Singh, and A. N. Yadav. 2021. Fungal enzymes: degradation and detoxification of organic and inorganic pollutants. In: Yadav A. N. (ed.), *Recent trends in mycological research: volume 2: Environmental and industrial perspective* (pp.99–125). Cham: Springer

Kuppusamy, S., P. Thavamani, K. Venkateswarlu, Y. B. Lee, R. Naidu, and M. Megharaj. 2017. Remediation approaches for polycyclic aromatic hydrocarbons (PAHs) contaminated soils: technological constraints, emerging trends and future directions. *Chemosphere.* 168: 944–968. https://doi.org/10.1016/j.chemosphere.2016.10.115

Lafortune, I., P. Juteau, E. Déziel, F. Lépine, R. Beaudet, and R. Villemur. 2009. Bacterial diversity of a consortium degrading high-molecular-weight polycyclic aromatic hydrocarbons in a two-liquid phase biosystem. *Microb. Ecol.* 57: 455–468. https://doi.org/10.1007/s00248-008-9417-4

Leahy, J. G., and R. R. Colwell. 1990. Microbial degradation of hydrocarbons in the environment. *Microbiol. Rev.* 54(3): 305–315.

Li, Y., X. Bai, Y. Ren, R. Gao, Y. Ji, Y. Wang, and H. Li. 2022. PAHs and nitro-PAHs in urban Beijing from 2017 to 2018: characteristics, sources, transformation mechanism and risk assessment. *J. Hazard. Mater.* 436: 129143. https://doi.org/10.1016/j.jhazmat.2022.129143

Luch, A., and W. M. Baird. 2005. Detoxification of polycyclic aromatic hydrocarbons. In: Luch, A., (ed.), *The carcinogenic effects of polycyclic aromatic hydrocarbons* (pp. 19–96). London: Imperial College Press.

Mandree, P., W. Masika, J. Naicker, G. Moonsamy, S. Ramchuran, and R. Lalloo. 2021. Bioremediation of polycyclic aromatic hydrocarbons from industry contaminated soil using indigenous bacillus spp. *Processes.* 9(9): 1606. https://doi.org/10.3390/pr9091606

Moubarz, G., A. Saad-Hussein, E. M. Shahy, H. Mahdy-Abdallah, A. M. Mohammed, L. A. Saleh, M. A. Abo-Zeid, and M. T. Abo-Elfadl. 2022. Lung cancer risk in workers occupationally exposed to polycyclic aromatic hydrocarbons with emphasis on the role of DNA repair gene. *Int. Arch. Occup. Environ. Health*. 1–17. https://doi.org/10.1007/s00420-022-01926-9

Mukherjee, P., and P. Roy. 2013. Copper enhanced monooxygenase activity and FT-IR spectroscopic characterisation of biotransformation products in trichloroethylene degrading bacterium: stenotrophomonas maltophilia PM102. *Biomed Res. Int*. https://doi.org/10.1155/2013/723680

National Center for Biotechnology Information. 2023. *PubChem compound summary for CID 9171*. Chrysene. Retrieved April 17, 2023 from https://pubchem.ncbi.nlm.nih.gov/compound/Chrysene.

Nayak, A. S., S. Sanjeev Kumar, M. Santosh Kumar, O. Anjaneya, and T. B. Karegoudar. 2011. A catabolic pathway for the degradation of chrysene by *Pseudoxanthomonas* sp. PNK-04. *FEMS Microbiol. Lett*. 320(2): 128–134.

Nwanna, I. M., G. O. George, and I. M. Olusoji. 2006. Growth study on chrysene degraders isolated from polycyclic aromatic hydrocarbon polluted soils in Nigeria. *Afr. J. Biotechnol*. 5(10): 823–828.

Nwinyi, O. C., F. W. Picardal, T. T. An, and O. O. Amund. 2013. *Aerobic degradation of naphthalene, fluoranthene, pyrene and chrysene using indigenous strains of bacteria isolated from a former Industrial site. Can. J. Pure Appl. Sci*. 7(2).

Nzila, A., and M. M. Musa. 2021. Current status of and future perspectives in bacterial degradation of benzo [a] pyrene. *Int. J. Environ. Res. Public Health*. 18(1): 262. https://doi.org/10.3934%2Fmicrobiol.2019.4.308

Olowomofe, T. O., J. O. Oluyege, B. I. Aderiye, and O. A. Oluwole. 2019. Degradation of poly aromatic fractions of crude oil and detection of catabolic genes in hydrocarbon-degrading bacteria isolated from Agbabu bitumen sediments in Ondo State. *AIMS Microbiol*. 5(4): 308. https://doi.org/10.3934%2Fmicrobiol.2019.4.308

Passarini, M. R., L. D. Sette, and M. V. Rodrigues. 2011. Improved extraction method to evaluate the degradation of selected PAHs by marine fungi grown in fermentative medium. *J. Braz. Chem. Soc*. 22: 564–570.

Patel, V., S. Cheturvedula, and D. Madamwar. 2012a. Phenanthrene degradation by Pseudoxanthomonas sp. DMVP2 isolated from hydrocarbon contaminated sediment of Amlakhadi canal, Gujarat, India. *J. Hazard. Mater*. 201: 43–51. https://doi.org/10.1016/j.jhazmat.2011.11.002

Patel, V., S. Jain, and D. Madamwar. 2012b. Naphthalene degradation by bacterial consortium (DV-AL) developed from Alang-Sosiya ship breaking yard, Gujarat, India. *Bioresour. Technol*. 107: 122–130. https://doi.org/10.1016/j.biortech.2011.12.056

Phillips, D. H., and Grover, P. L. 1994. Polycyclic hydrocarbon activation: bay regions and beyond. *Drug Metab. Rev*. 26(1–2): 443–467.

Pinedo-Rivilla, C., J. Aleu, and I. G. Collado. 2009. Pollutants biodegradation by fungi. *Curr. Org. Chem*. 13(12): 1194–1214. https://doi.org/10.2174/138527209788921774

Potin, O., E. Veignie, and C. Rafin. 2004. Biodegradation of polycyclic aromatic hydrocarbons (PAHs) by Cladosporium sphaerospermum isolated from an aged PAH contaminated soil. *FEMS Microbiol. Ecol*. 51(1): 71–78.

Pramanik, K., S. Kundu, S. Banerjee, P. K. Ghosh, and T. K. Maiti. 2018. Computational-based structural, functional and phylogenetic analysis of *Enterobacter* phytases. *3 Biotech*. 8: 1–12. https://doi.org/10.1007/s13205-018-1287-y

Premnath, N., K. Mohanrasu, R. G. R. Rao, G. H. Dinesh, G. S. Prakash, V. Ananthi,... A. Arun. 2021. A crucial review on polycyclic aromatic hydrocarbons-environmental occurrence and strategies for microbial degradation. *Chemosphere*. 280: 130608. https://doi.org/10.1016/j.chemosphere.2021.130608

Rafeeq, H., N. Afsheen, S. Rafique, A. Arshad, M. Intisar, A. Hussain, M. Bilal, and H. M. Iqbal. 2023. Genetically engineered microorganisms for environmental remediation. *Chemosphere.* 310: 136751. https://doi.org/10.1016/j.chemosphere.2022.136751

Raghukumar, C., M. S. Shailaja, P. S. Parameswaran, and S. Singh. 2006. *Removal of polycyclic aromatic hydrocarbons from aqueous media by the marine fungus NIOCC 312: involvement of lignin-degrading enzymes and exopolysaccharides. Indian J. Mar. Sci.* 35(4): 373–379.

Ravanbakhsh, M., H. Yousefi, E. Lak, M. J. Ansari, W. Suksatan, Q. A. Qasim, ... M. J. Mohammadi. 2022. Effect of polycyclic aromatic hydrocarbons (PAHs) on respiratory diseases and the risk factors related to cancer. Polycyclic Aromatic Compounds 43(9): 8371–8387.. https://doi.org/10.1080/10406638.2022.2149569

Riskuwa-Shehu, M. L., H. Y. Ismail, A. Y. Fardami, and U. B. Ibrahim. 2022. Pennisetum purpureum improved polycyclic aromatic hydrocarbons removal in weathered-petroleum contaminated soil. *European J. Biol. Biotechnol.* 3(3): 7–13. https://doi.org/10.24018/ejbio.2022.3.3.285

Roszak, M., J. Jabłońska, X. Stachurska, K. Dubrowska, J. Kajdanowicz, M. Gołębiewska,... J. Karakulska. 2021. Development of an autochthonous microbial consortium for enhanced bioremediation of PAH-contaminated soil. *Int. J. Mol Sci.* 22(24): 13469. https://doi.org/10.3390/ijms222413469

Saiu, G., S. Tronci, M. Grosso, E. Cadoni, and N. Curreli. 2016. Biodegradation of polycyclic aromatic hydrocarbons by pleurotus sajor-caju. *Chem. Eng. Trans.* 49: 487–492.

Seo, J. S., Y. S. Keum, and Q. X. Li. 2009. Bacterial degradation of aromatic compounds. *Int. J. Environ. Res. Public Health* 6(1): 278–309.

Show, B. K., S. Banerjee, A. Banerjee, R. GhoshThakur, A. K. Hazra, N. C. Mandal, A. B. Ross, S. Balachandran, and S. Chaudhury. 2022. Insect gut bacteria: a promising tool for enhanced biogas production. *Rev. Environ. Sci. Biotechnol.* 21(1): 1–25. https://doi.org/10.1007/s11157-021-09607-8

Silva, I. S., M. Grossman, and L. R. Durran. 2009. Degradation of polycyclic aromatic hydrocarbons (2–7 rings) under microaerobic and very-low-oxygen conditions by soil fungi. *J. Bioremediat. Biodegrad.* 63(2): 224–229. https://doi.org/10.1016/j.ibiod.2008.09.008

Singh, A., and O. P. Ward. 2004. Applied bioremediation and phytoremediation (series: soil biology, vol 1). *J. Soils Sediments.* 4(3), 209.

Smith, J. R., D. V. Nakles, D. F. Sherman, E. F. Neuhauser, and R. C. Loehr. 1989. Environmental fate mechanisms influencing biological degradation of coal-tar derived polynuclear aromatic hydrocarbons in soil systems. In: Lyne D., (ed.), *The third international conference on new frontiers for hazardous waste management* (pp. 397–405). Washington, DC: US Environmental Protection Agency.

Tao, L. P., X. Li, M. Z. Zhao, J. R. Shi, S. Q. Ji, W. Y. Jiang,... S. Y. Zhang. 2021. Chrysene, a four-ring polycyclic aromatic hydrocarbon, induces hepatotoxicity in mice by activation of the aryl hydrocarbon receptor (AhR). *Chemosphere.* 276. 130108. https://doi.org/10.1016/j.chemosphere.2021.130108

Tavares, A. P. M., M. A. Z. Coelho, M. S. M. Agapito, J. A. P. Coutinho, and A. M. R. B. Xavier. 2006. Optimisation and modeling of laccase production by Trametes versicolor in a bioreactor using statistical experimental design. *Appl. Biochem. Biotechnol.* 134: 233–248. https://doi.org/10.1385/ABAB:134:3:233

Thacharodi, A., S. Hassan, T. Singh, R. Mandal, H. A. Khan, M. A. Hussain, and A. Pugazhendhi. 2023. Bioremediation of polycyclic aromatic hydrocarbons: an updated microbiological review. *Chemosphere.* 138498. https://doi.org/10.1016/j.chemosphere.2023.138498

Thomas, S., N. T. Veettil, and K. Subbiah. 2021. Isolation, characterisation and optimisation of chrysene degradation using bacteria isolated from oil-contaminated water. *Water Sci. Technol.* 84(10–11): 2737–2748. https://doi.org/10.2166/wst.2021.227

USEPA. 2002. *Polycyclic Organic Matter*. US Environmental Protection Agency. (https://www.epa.gov/ttn/atw/hlthef/polycycl.html). Washington, D.C.

Vaidya, S., N. Devpura, K. Jain, and D. Madamwar. 2018. Degradation of chrysene by enriched bacterial consortium. *Front. Microbiol.* 9: 1333. https://doi.org/10.3389/fmicb.2018.01333

Verstraete, W., R. Vanloocke, R. DeBorger, and A. Verlinde. 1976. Modelling of the breakdown and the mobilisation of hydrocarbons in unsaturated soil layers. In: *Proceedings of the 3rd International Biodegradation Symposium* (pp. 99–112). London: Applied Science Publishers Ltd.

Vipotnik, Z., M. Michelin, and T. Tavares. 2021. Ligninolytic enzymes production during polycyclic aromatic hydrocarbons degradation: effect of soil pH, soil amendments and fungal co-cultivation. *Biodegradation*. 32: 193–215. https://doi.org/10.1007/s10532-021-09933-2

Vipotnik, Z., M. Michelin, and T. Tavares. 2022. Biodegradation of chrysene and benzo [a] pyrene and removal of metals from naturally contaminated soil by isolated Trametes versicolor strain and laccase produced thereof. *Environ. Technol. Innov.* 28: 102737. https://doi.org/10.1016/j.eti.2022.102737

Vondráček, J., and M. Machala. 2021. The role of metabolism in toxicity of polycyclic aromatic hydrocarbons and their non-genotoxic modes of action. *Curr. Drug Metab.* 22(8): 584–595. https://doi.org/10.2174/1389200221999201125205725

Willison, J. C. 2004. Isolation and characterisation of a novel sphingomonad capable of growth with chrysene as sole carbon and energy source. *FEMS Microbiol. Lett.* 241(2): 143–150.

Yamini, V. and V. D. Rajeswari. 2023. Metabolic capacity to alter polycyclic aromatic hydrocarbons and its microbe-mediated remediation. *Chemosphere.* 138707. https://doi.org/10.1016/j.chemosphere.2023.138707

Yang, B., Y. Shi, S. Xu, Y. Wang, S. Kong, Z. Cai, and J. Wang. 2022. Polycyclic aromatic hydrocarbon occurrence in forest soils in response to fires: a summary across sites. *Environ. Sci. Process. Impacts.* 24(1): 32–41.

Zheng, Z., and J. P. Obbard. 2003. Oxidation of polycyclic aromatic hydrocarbons by fungal isolates from an oil contaminated refinery soil. *Environ. Sci. Pollut. Res.* 10: 173–176. https://doi.org/10.1065/espr2002.07.126

Zhou, Q., R. Li, T. Li, R. Zhou, Z. Hou, and X. Zhang. 2023. Interactions among microorganisms functionally active for electron transfer and pollutant degradation in natural environments. *Eco-Environ Health.* 2: 3–15. https://doi.org/10.1016/j.eehl.2023.01.002

Zimmerman, A. R., D. H. Kang, M. Y. Ahn, S. Hyun and M. K. Banks. 2008. Influence of a soil enzyme on iron-cyanide complex speciation and mineral adsorption. *Chemosphere.* 70(6): 1044–1051. https://doi.org/10.1016/j.chemosphere.2007.07.075

4 Environmental Remediation of Noxious Agrochemicals Using Engineered Materials of Bio-Origin

Manvendra Patel, Sarita Dhaka,
Anuj Rana, and Rahul Kumar Dhaka

4.1 INTRODUCTION

Modern agriculture is heavily dependent upon a variety of inputs in the form of chemicals called agrochemicals. These chemicals are extensively utilized globally for enhancing the growth of agricultural produce to feed the people. Agrochemicals are also used in horticulture, forestry, pest control, and weed control. Agrochemicals include both organic and inorganic compounds. Commonly, the term agrochemical refers to the group of pesticides which include insecticides, fungicides, herbicides, rodenticides, nematicides, and molluscicides. But agrochemicals also include fertilizers and soil conditioners such as biosolids, liming, and acidifying agents. The groups of fertilizers and pesticides comprise the largest portion of agrochemicals.

Agrochemicals demand also increased rapidly with the rise in human population and food demand in the past few decades. Globally, 3 billion kg pesticides were used annually (Sharma et al., 2020). The global fertilizer (ammonia, phosphoric acid, and potash) supply for 2020 was >273 million tons (FAO, 2017). The presence of these pesticides and fertilizers in the aquatic and terrestrial environments poses a great threat to biogeochemical processes, flora, fauna, and humans (Sharma et al., 2020). A joint report by United Nations Education programme (UNEP) and World health organization (WHO) estimates 200,000 human deaths and 3 million pesticide poisoning cases globally each year (Md Meftaul et al., 2020). Thus, the elimination of agrochemicals from environmental matrices becomes essential.

Remediation of agrochemicals from environmental systems is challenging. In soil, the agrochemical immobilization or elimination through low-cost materials is the main challenge. While in aquatic sources, the main challenge is posed due to extremely low concentrations and wide variations in aqueous matrix. Various technologies, including reverse osmosis, nano-filtration, coagulation, biodegradation, photolysis, advanced oxidation processes (AOPs), and adsorption, were evaluated for

DOI: 10.1201/9781003407317-4

a wide range of contaminants, including agrochemicals (Singh et al., 2020). Except adsorption, most of these techniques have several limitations, including sludge production, toxic metabolite generation, low removal efficiencies, high implementation and operation cost, and the requirement of skilled personnel for operating the treatment facilities (Jain et al., 2022, Patel et al., 2019, Pratap et al., 2021). Thus, developing engineered materials of bio-origin with multiple properties of contaminants sorption, immobilization, and degradation can be the ultimate solution, especially in developing countries, including India.

Pesticides and herbicides are the most detected agrochemicals with high toxicity as compared to fertilizers. Thus, this chapter focuses on pesticides and herbicides remediation from aquatic and soil systems using materials of bio-origin. This chapter summarizes the importance of utilizing materials of bio-origin for agrochemical remediation from environmental systems. This chapter provides the development details of various bio-origin materials, their properties, and application for agrochemical remediation from soil and aquatic systems. The advantages of utilizing materials of bio-origin, including benefits of bio-origin waste management, contaminants elimination, and environmental sustainability in a low-cost manner, are discussed. A further mechanistic aspect of agrochemical mitigation is also discussed.

4.2   ENVIRONMENTAL OCCURRENCE OF AGROCHEMICALS

Agrochemical applications for agriculture, horticulture, and gardens release both organic and inorganic contaminants. Common agrochemicals release heavy metals, inorganic ions, nutrients, and synthetic organic compounds. Recently, a variety of emerging contaminants and microplastics have also been reported in agricultural fields, gardens, and aquatic systems due to the application of plastic mulches and biosolids (Clarke and Smith, 2011, Ng et al., 2018). Several agrochemicals have been previously reported in aquatic and soil systems around the world (Sharma et al., 2020, Md Meftaul et al., 2020).

Pesticides are one of the most common agrochemicals in soil and are consistently detected in aqueous systems too. In an extensive study between 2013 and 2017, the USGS (United States Geological Survey) found the presence of 221 different pesticides from 72 river and stream sites, with 88% of samples having five or more pesticides (Covert et al., 2020). A nationwide study of tap water in 38 cities in China showed at least one neonicotinoid in every sample from the cities (He et al., 2021). Pesticide pollution is also increasing in developing countries, including India, Nepal, Pakistan, and Bangladesh (Md Meftaul et al., 2020).

Compared to developed countries, pesticide and herbicide occurrence studies are limited. Still, both pesticides and herbicides were detected in soil and aquatic systems in India previously (Yadav et al., 2015). One of the earliest studies from 1989 reported the presence of hexachlorocyllohexane (HCH) and Dichlorodiphenyl trichloroethylene (DDT) from wells, hand pumps, and ponds in Bhopal, with well water containing 4640 and 5794 µg/L HCH and DDT, respectively (Bouwer, 1989, Yadav et al., 2015). Another study detected HCH (23.9–2488 ng/L) and DDT (10.9–314.9 ng/L) in drinking water from Ahmedabad (Jani et al., 1991). Further pesticides were reported in groundwaters from Aligarh (Ray, 1989), Bhandra, Amravati, Yavatmal

(Lari et al., 2014), Ambala, Gurgaon (Kaushik et al., 2012),, and many other locations in India (Yadav et al., 2015). Pesticides are also reported from several Indian rivers, including Ganges (Mutiyar and Mittal, 2013), Yamuna (Kaushik et al., 2008), Tamiraparani (Kumarasamy et al., 2012), Gomti (Malik et al., 2009), and Cauvery (Abida et al., 2009). Similarly, pesticide residues were reported from several surface water and soil systems (Yadav et al., 2015, Ashesh et al., 2022, Kaushik et al., 2012). Altogether, these studies validate the extensive presence of pesticides in Indian surface waters, groundwaters, and soil systems.

4.3 MATERIALS OF BIO-ORIGIN

Pristine materials of natural/biological origin and their modified forms are the materials of bio-origin. These materials are provided by microorganisms, plants, and animals. Alginate, biochar, biomass, cellulose, chitin, chitosan, and peat are some of the materials of bio-origin that are extensively applied for contaminants removal. These materials of bio-origin are used directly or can be modified before their application. Common modification techniques are size reduction, pyrolysis, activation, impregnation, magnetization, and composite development (Pratap et al., 2022, Pratap et al., 2021, Krasucka et al., 2021, Jain et al., 2022, Singh et al., 2020). Modification or engineering approaches provide several advantages, including enhanced porosity, functionality, and higher and selective sorption (Pratap et al., 2022, Pratap et al., 2021, Krasucka et al., 2021, Jain et al., 2022, Singh et al., 2020).

4.4 REMOVAL OF AGROCHEMICALS FROM AQUEOUS MEDIA

The presence of agrochemicals in surface water, groundwater, and wastewater is well established. Thus, utilizing low-cost engineered materials of bio-origin for remediations of the noxious agrochemicals from the contaminated aqueous and soil systems can be a sustainable solution. Microorganisms, plants, and animals produce the materials of bio-origin. Commonly used materials of bio-origin are alginate, biochar, biomass, cellulose, chitin, chitosan, and peat. Bio-origin materials offer different characteristics like natural abundance, cost-effectiveness, biodegradability, biocompatibility, and less energy consumption in contrast to synthetic materials (Singh et al., 2020). Moreover, engineered biomaterials are tuned to have a high affinity for the removal of agrochemicals from the aqueous and soil systems.

4.4.1 REMOVAL OF AGROCHEMICALS VIA ADSORPTION

Adsorption is the attachment of fluid phase molecules to the solid phase (Patel et al., 2019). Adsorption is a simple, cost-effective, and efficient solution for the decontamination of water from both organic and inorganic contaminants (Jain et al., 2022). Adsorption is performed as batch and column studies. Adsorption also has the advantage of application in a wide range of aqueous solutions from drinking water to wastewater (Ateia et al., 2020). The advantages of adsorption over other techniques include

- No secondary pollutant or sludge generation,
- Low cost,
- Wide applicability, and
- Reusability.

The important factors affecting sorption include the pH of the solution, particle size of the adsorbent, affinity of the solute toward sorbent, and contact time (Patel et al., 2019). An understanding of these factors is necessary for determining the sorption mechanism (Patel et al., 2019). The solution pH affects the sorption by altering the speciation of adsorbate with changing pH (Patel et al., 2021, 2022a, Vimal et al., 2019). For example, carbofuran exists as protonated, neutral, and deprotonated (Figure 4.1). Solution pH also affects the pH point of zero charge (pH_{pzc}) of the adsorbent.

Any change in the solution's matrix, i.e., presence of different ions and organic matter, can affect the sorption of contaminants, including agrochemicals(Cao et al., 2021, Lee et al., 2021). For example, sorption of atrazine on fallen leaf biochar and nano-MgO-modified fallen, leaf biochar is affected by change in solution NaCl, $CaCl_2$, and humic acid concentration (Cao et al., 2021). Similarly, in another study, ionic strength significantly affected the sorption of alachlor, diuron, and simazine showing a significant effect of ionic strength on ground coffee biochar and NaOH-activated ground coffee biochar (Lee et al., 2021).

Utilizing pristine and engineered materials of bio-origin had added advantages of low treatment cost, waste management, local availability, and ease of application (Jain et al., 2022, Singh et al., 2020, Pratap et al., 2021, 2022). Alginate, biomass, cellulose, chitin, chitosan, and peat are the common materials of bio-origin that can be used directly or engineered to develop adsorbents with high removal capacity. Several bio-origin materials had been previously used for agrochemical sorption (Gautam et al., 2021, Jain et al., 2022, Singh et al., 2020).

4.4.1.1 Chitosan-Based Adsorbents

Chitin and chitosan are white, hard inelastic nitrogenous, natural polysaccharide biopolymers. They serve as supporting material for insects and crustaceans (Singh et al., 2020). Chitosan (poly[β-(1–4)-2-amino-2-deoxy-D-glucopyranose]) is a cationic natural, cellulosic biopolymer with high molecular weight. It is commercially obtained through deacetylation of chitin by thermochemical reaction (Moradi Dehaghi et al., 2014). The N-acetyl-glucosamine content differentiates them, with chitin having

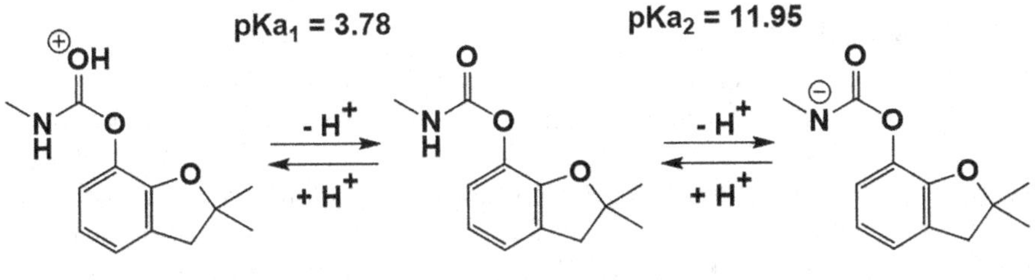

FIGURE 4.1 Speciation of carbofuran with change in solution pH.

<50% units, while chitosan having >50% biopolymer units. Chitin and chitosan are readily available in large amounts as a by-product from food processing industries. Chitosan and chitin are efficient biosorbents for many contaminants, including agrochemicals. Chitosan offers several advantages, including non-toxicity, non-allergic, low cost, and hydrophilicity, with the ability to be modified into materials with enhanced and specific properties (Moradi Dehaghi et al., 2014).

Chemical modification of chitin and chitosan enhances the selectivity and sorption capacity and reduces the solubility problem (Moradi Dehaghi et al., 2014, Ranjbar Bandforuzi and Hadjmohammadi, 2019, Singh et al., 2020). For example, mixed hemimicelle SDS-coated magnetic chitosan nanoparticles (MHMS-MCNPs) were developed and applied for the preconcentration and removal of chlorpyrifos, diazinon, and phosalone (Ranjbar Bandforuzi and Hadjmohammadi, 2019). MHMS-MCNPs were developed through an in situ Fe(II) and Fe(III) co-precipitation method in the presence of chitosan and sodium tripolyphosphate. MHMS-MCNPs showed 13.48, 16.58, and 15.53 mg/g Langmuir capacities for chlorpyrifos, diazinon, and phosalone (Ranjbar Bandforuzi and Hadjmohammadi, 2019). In another study, chitosan-ZnO nanoparticles (CS-ZnONPs) composite beads were developed and utilized for permethrin sorption (Moradi Dehaghi et al., 2014). CS-ZnONPs biocomposite was developed through a simple method where 1 g chitosan was added to a 10% nitric acid solution in 1% acetic acid. The mixture was sonicated, and NaOH solution was added drop by drop to maintain a continuous pH 10 and kept at 60°C for 3 hours. The modification not only enhanced the permethrin sorptive property (from 49% to 99%) but also reduced the chitosan sweeling from 35.2% to 13.7% for CS-ZnONPs (Moradi Dehaghi et al., 2014).

Magnetic chitosan graphene oxide composite (GO-Chm) was prepared and applied for the sorption of isoproturon, linuron, and monuron herbicides (Shah et al., 2018). GO-Chm showed Langmuir capacities of 29.4, 33.3, and 35.7 mg/g for isoproturon, linuron, and monuron, respectively. Furthermore, GO-Chm was also utilized for the extraction of the herbicides from aqueous systems and rice for analysis (Shah et al., 2018). While another low-cost chitosan-calcite adsorbent was developed using shrimp residue for triazines (prometryn, metribuzin, and terbuthylazine) removal (Borja-Urzola et al., 2021). The adsorption studies were performed in a batch system to study the effects of adsorbent dose, initial triazine concentration, pH of the solution, and contact time. The study showed the role of hydrogen bonding and hydrophobic and dipole–dipole interactions in the sorption of prometryn, metribuzin, and terbuthylazine (Borja-Urzola et al., 2021). Together, all these studies showed the effectiveness of chitosan and modified chitosan in agrochemicals sorption.

4.4.1.2 Biosorbents

Biosorbents are living or dead biomass, i.e., materials of biological origin. They include algae, microbes (including fungi, yeast, bacteria), agricultural residues (bran, husk, hulls, straws, and woody biomass), forest products (wood, leaf litter bark, and peat), food wastes (fruit peels, vegetable residues, pulp wastes, etc.), and industrial wastes (sludge and fermentation wastes) and other materials such as chitin and chitosan (Singh et al., 2020, Okoro et al., 2022). These materials can be utilized without modification or after modification as well (Schwantes et al., 2022). Utilizing these materials has another advantage of biomass-based solid waste management. Some of the specific biosorbents such as cellulose and chitosan are discussed separately in the sections below.

In a study, cassava peel, crambe meal, and pinus bark biomasses were successfully utilized for chlorpyrifos removal with and without modification (Schwantes et al., 2022). Of the biomasses, the crambe meal showed the highest chlorpyrifos removal. The biomasses were modified using H_2O_2, NaOH, and H_2SO_4, and H_2O_2 modification showed the highest removal for all three biomasses (Schwantes et al., 2022). The study reported the role of hydrogen bonding, π-π interactions and electrostatic and hydrophobic interactions (Schwantes et al., 2022). In another study, chickpea husk was used for the sorption of organophosphorus pesticides (triazophos and methyl parathion) from aqueous solution (Akhtar et al., 2009). The Langmuir capacities of triazophos and methyl parathion were 0.0077 and 0.025 mmol/g, respectively (Akhtar et al., 2009). In another study, five selected pesticides (atrazine, chloropyrifos, fluzifop-P-butyl, lambda-cyhalothrin, and lactofen) on cork species (*Quercus ceris* and *Quercus suber*) tree derived biosorbents (de Aguiar et al., 2019). The study showed pesticide removal efficiencies between 70% and 80% for both biosorbents with the highest removal at pH 3.0°C and 30°C temperature (de Aguiar et al., 2019).

Biosorbents pose some disadvantages, including fouling, less stability, and low sorption capacity (Jain et al., 2022, Guo et al., 2022). These problems can be mitigated with the development of modified biosorbents (Guo et al., 2022). For example, sugarcane bagasse was modified using tetraethylenepentamine (TEPA) to develop amine-modified bagasse for glyphosate and other contaminants sorption (Guo et al., 2022). The amine-modified bagasse showed excellent sorption potential with a Langmuir capacity of 227.4 mg/g for glyphosate (Guo et al., 2022). Industrial waste biomasses contain various elements, which can be utilized for the enhanced adsorption of certain contaminants. For example, tannery waste sludge was used for the development of multimetal-loaded biosorbent and utilized for glyphosate sorption (Ramrakhiani et al., 2019). The multimetal-loaded biosorbent was developed by adsorbing Ni(II), Cd(II), Co(II), and Zn(II) on tannery waste sludge. The multi-metal loaded tannery waste sludge biosorbent showed an increase in glyphosate sorption efficiency from 26% to >92% (Ramrakhiani et al., 2019). The study showed that the metals play a crucial role in glyphosate removal through phosphate-metal complex formations (Ramrakhiani et al., 2019).

4.4.1.3 Cellulose-Based Adsorbents

Cellulose ($C_6H_{10}O_5$)n is the prime constituent of biomass and is one of the most abundant natural materials with an approximate production of 10^{11} to 10^{12} tons/year (Suhas et al., 2016). Wood and cotton are the commercial sources of cellulose. Cellulose is a biodegradable sorbent with high hydrophilicity, non-toxic nature, and modification potential for enhanced properties. Cellulose is a linear homopolymer made from D-anhydroglucopyranose monomer units. A variety of cellulose-based adsorbents, including nanocellulose and cellulose fibers, were developed for pesticide sorption (Rana et al., 2021). Pristine cellulose-based adsorbents showed limited sorption potential (Rana et al., 2021). Modification of cellulose-based adsorbents can further enhance the sorptive properties and stability (Hokkanen et al., 2016).

Cellulose can be modified to develop cellulose/clay nanocomposites, cellulose/ graphene nanocomposites, cellulose/metal oxide nanocomposites, magnetic cellulose nanocomposites, etc. (Rana et al., 2021).

For example, the biopolymer-nano-organoclay composite beads on carboxy methyl cellulose were developed for nine pesticides (butachlor, atrazine, carbofuran, carbendazim, isoproturon, imidacloprid, pendimethalin, thiamethoxam, and thiophanate methyl) removal (Narayanan et al., 2020). The study reported a removal efficiency of 57%–100% in water and 63%–91% in industrial effluents (Narayanan et al., 2020). The authors also showed the high recyclability potential by regenerating the developed composite using acetone washing followed by the thermal regeneration at 200°C. The regenerated composite maintains 57%–97% removal efficiency even after second generation (Narayanan et al., 2020). In another study, graphene oxide/ electrospun cellulose nanofiber (GO/CNF-10%) was developed for organophosphorus pesticide sorption and further utilized for their extraction and analysis from environmental samples (Aris et al., 2020). The developed adsorbent (GO/CNF-10%) proved to be an inexpensive and efficient material for organophosphorus pesticide (chloropyrifos, ethoprophos, methyl parathion, and sulfotepp) adsorption.

Metal oxide nanoparticle cellulose composites provide several advantages including enhanced sorption, enhanced stability, and photocatalytic properties (Rana et al., 2021). For example, Cu-MOF@cellulose acetate (Cu-BTC@CA) membrane with a high surface area (BET surface area = $965.8 \, m^2/g$) was developed for the adsorption of dimethoate (Abdelhameed et al., 2021). The Cu-BTC@CA showed enhanced dimethoate sorption (282.3–321.9 mg/g) versus cellulose acetate membrane (207.8 mg/g). The Cu-BTC@CA also showed high recyclability as only 22.5% reduction in adsorption efficiency was observed even after five regeneration cycles (Abdelhameed et al., 2021). In another study, cadmium sulfide nanoparticles were encrusted into biomass-derived silanized cellulose nanofibers for adsorptive pesticide detoxification (Komal et al., 2020). The developed adsorbent CdS@10%SCNF showed excellent chlorpyrifos sorption with a Langmuir capacity of 86.96 mg/g. Furthermore, CdS@10%SCNF also showed excellent regenerative and recyclability potential (Komal et al., 2020).

4.4.1.4 Alginate-Based Adsorbents

Alginate is a natural water-soluble linear polysaccharide derived from seaweeds (Yu et al., 2017). Adsorbents developed from alginate have fascinating features, including biodegradable, non-toxic, biocompatibility, high recyclability, renewability, tuneable physicochemical properties, and excellent sorption potential (Qamar et al., 2022, Sutirman et al., 2021). Alginate can be processed into hydrogels and other adsorbents after a reaction with multivalent metal ions such as calcium (Singh et al., 2020). Alginate membranes were developed for herbicides (difenzoquat and diquat) sorption (Agostini de Moraes et al., 2013). The membrane showed 62% and 95% adsorption for difenzoquat and diquat, respectively, in 120 minutes at an initial concentrations of 50 µM. The Langmuir capacity of difenzoquat was 34.19 mg/g, and electrostatic interactions play an important role in herbicide sorption on alginate membrane (Agostini de Moraes et al., 2013).

Alginate can be used to develop a variety of adsorbents, including beads, hydrogels, nanocomposites, and composites with different materials such as biochar, clay, and starch (Aziz et al., 2023, Jacob et al., 2022, Du et al., 2023, Wang et al., 2022, Zhou et al., 2022). For example, in a study, adsorptive decontamination (86% under optimal conditions) of chlorpyrifos was successfully achieved through sugarcane bagasse-derived biochar-alginate beads (Jacob et al., 2022). The biochar-alginate beads were prepared at 3:1 sodium alginate to biochar ratio and showed a maximum adsorption capacity of 6.25 mg/g for chlorpyrifos (Jacob et al., 2022). In another study, copper doped *Cerastoderma edule* shells and alginate biocomposite hydrogel beads were developed for fungicide thiabendazole removal (Aziz et al., 2023). The hydrogel beads showed a maximum Langmuir capacity of 21.89 mg/g with pseudo-second-order showing better fitting to the sorption dynamics (Aziz et al., 2023). Similarly, calcium alginate beads were developed and utilized for difenoconazole and nitenpyram removal (Zhou et al., 2022).

4.4.1.5   Carbonaceous Adsorbents

Several carbonaceous adsorbents, including peat, biochar, engineered biochars, and activated carbons, have been directly derived from biomass. These adsorbents have also been extensively applied for agrochemical sorption from aqueous solutions (Gautam et al., 2021, Essandoh et al., 2017a, b, Mayakaduwa et al., 2017, Pratap et al., 2021, 2022, Vimal et al., 2019, Vithanage et al., 2016).

Biochar is a multifunctional carbonaceous material produced by the thermochemical conversion of biomass (Ahmad et al., 2014, Mohan et al., 2014). In the past 15 years, biochars have been seen as an alternative to activated carbon in environmental applications. High surface area, aromaticity, porosity, and abundant functionality make biochar an effective adsorbent for a variety of contaminants sorption, including agrochemicals (Ahmad et al., 2014, Essandoh et al., 2017a, b, Mayakaduwa et al., 2017, Mohan et al., 2014, Patel et al., 2019, 2021, 2022b, Vimal et al., 2019). Furthermore, biochar can also be engineered to provide selective and enhanced agrochemicals removal (Pratap et al., 2021, 2022, Gautam et al., 2021).

Sugarcane bagasse biochar developed at 500°C was applied for aqueous carbofuran sorption (Vimal et al., 2019). The study showed maximum carbofuran sorption at pH 2.0, with a maximum sorption capacity of 18.91 mg/g at 45°C (Vimal et al., 2019). Ground coffee residue biochar (GCRB) developed at 800°C was applied for the sorption of alachlor, diuron, and simazine herbicides. The GCRB showed adsorption capacities of 11.74, 9.95, and 6.53 µmol/g for alachlor, diuron, and simazine, respectively, with inner layer complexation playing a critical role in the sorption (Lee et al., 2021). In another study, clomazone and cyhalofop acid were adsorbed onto grape wood pruning biochar prepared at different temperatures (350°C, 500°C, and 900°C). The highest clomazone and cyhalofop acid sorption was achieved for the biochar prepared at 900°C (Gámiz et al., 2019). The rise in organic contaminants sorption with a rise in biochar production temperature is a commonly reported trend (Patel et al., 2019, 2022a, Ahmad et al., 2014, Mohan et al., 2014, Gámiz et al., 2019). This is attributed to an increase in specific surface area and aromaticity with an increase in production temperature (Gámiz et al., 2019).

Modification or engineering approaches can be utilized to prepare biochars with enhanced contaminants removal capacity and enhanced properties (Krasucka et al., 2021, Patel et al., 2022b, Gámiz et al., 2019, Lee et al., 2021). These approaches include activation, metal impregnation, magnetization, and composites preparation (Krasucka et al., 2021). For example, the specific surface area and pore volume of GCRB increases 106 and 21 times after NaOH activation (Lee et al., 2021). Furthermore, NaOH activation of GCRB also increased the adsorption capacities from 11.74 to 122.71 µmol/g for alachlor, 9.95–166.42 µmol/g for diuron, and 6.53–99.16 µmol/g for simazine (Lee et al., 2021). Similarly, H_2O_2 activation of grape wood pruning biochar prepared at 350°C increased the removal efficiency from 6.3% to 35.4% for cyhalofop acid (Gámiz et al., 2019).

Biochar can be combined with various components, including clays, metal oxides, hydroxides, CNTs, graphene oxides, graphene, and polymers, to develop biochar composites (Krasucka et al., 2021). The composite development provides enhanced surface functionality, enhanced porosity, and enhanced stability along with enhanced contaminant removal capacity (Krasucka et al., 2021). The modification of fallen leaf biochar to develop nano-MgO/biochar composite increased atrazine sorption by 1.99–5.71 times with a maximum capacity of 22.4 mg/g (Cao et al., 2021). The nano-MgO biochar composite showed enhanced hydrophilicity and good reusability after five adsorption-desorption cycles (Cao et al., 2021). In another study, zerovalent iron supported on biochar nanocomposites (Fe^0-B_{RtP}) was developed from Rambutan fruit peel waste for organochlorine pesticide removal (Batool et al., 2022). The biochar nanocomposite was synthesized through a green method utilizing Rambutan peel extract as a green reducing agent. The biochar nanocomposite synthesized showed 89%–92% pesticide removal as compared to 5%–13% chemically synthesized biochar nanocomposite (Batool et al., 2022).

Peat is an abundant, readily available, and inexpensive biosorbent. It is a porous carbonaceous material containing a large amount of cellulose, lignin, and humic acid (Singh et al., 2020). The lignin and humic acid fractions of peat have polar functional groups, including alcohols, aldehydes, carboxyl, ketonic, phenolic, and ethers (Singh et al., 2020). Peat is utilized in biomixtures for contaminants removal, including agrochemicals. For example, successful atrazine removal was performed through a fixed-bed column packed with a biomixture of wheat straw, soil, and peat achieving the highest bed capacity of 95.32 mg/g (Levio-Raiman et al., 2021).

Peat as an adsorbent has demonstrated limited potential for aqueous contaminants removal due to poor chemical stability, high hydrophilicity, low mechanical strength, leaching of humic and fulvic acids, and shrinking and swelling tendency (Singh et al., 2020). However, physical and chemical modification improves sorption capacities and selectivity as well as eliminates the above-mentioned drawbacks (Singh et al., 2020). Pyrolysis, carbonization, and gasification are common physical methods for the conversion of peat biomass into carbonaceous adsorbents (Wiśniewska and Nowicki, 2020, Aroguz, 2006). For example, pyrolyzed ocean peat moss was developed at 600°C and utilized for azinphos-methyl insecticide sorption. The pyrolyzed peat moss showed a Langmuir capacity of 4.54 mg/g at 30°C (Aroguz, 2006).

4.4.2 MECHANISM INVOLVED IN THE SORPTION OF AGROCHEMICALS IN WATER

Accumulation of adsorbate molecules onto the surface of the adsorbent is adsorption. Adsorption is facilitated by the presence of different functional groups, minerals on the surface of adsorbent, and the porosity of the adsorbent (Patel et al., 2019, 2022a, Jain et al., 2022). Adsorption is also dependent on the physico-chemical properties of the adsorbate (Patel et al., 2019). Together, these properties determine the type of physical or chemical interaction playing the main role in the sorption of a particular adsorbate on an adsorbent (Patel et al., 2022a). Furthermore, reaction temperature and solution parameters, including pH, adsorbent dose, adsorbate concentration, contact time, ionic strength, anions, cations, and humic acid, also affect the sorption mechanism (Patel et al., 2021, Choudhary et al., 2020). Common mechanisms include hydrophobic interaction, H-bonding, electrostatic attraction, π-π interactions, ion exchange, precipitation, surface complexations, and pore diffusion (Ahmad et al., 2014, Mohan et al., 2014, Patel et al., 2019). Different mechanisms reported for the sorption of agrochemicals on materials of bio-origin are depicted in Figure 4.2. Further modification of adsorbents with metals can alter sorption mechanisms by introducing metal complexation interactions such as monodentate and bidentate complexations (Figure 4.3).

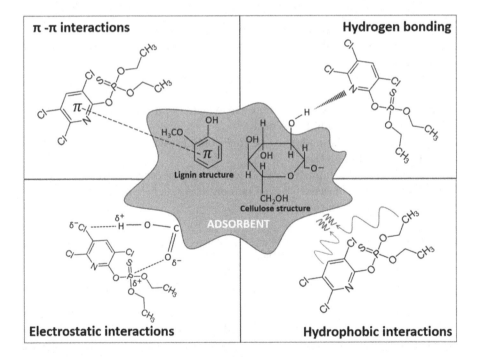

FIGURE 4.2 Schematic representation of possible chlorpyrifos adsorption mechanisms on biosorbent (Schwantes et al., 2022).

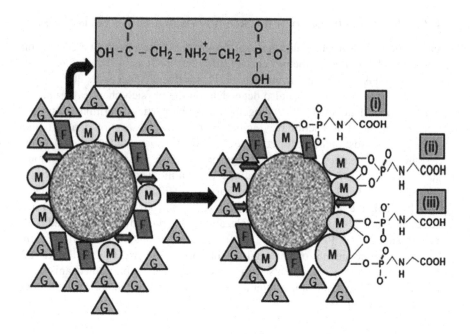

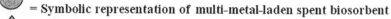

= Symbolic representation of multi-metal-laden spent biosorbent

= Glyphosate molecule in aqueous solution and its ionic form is denoted as inset

= Various functional groups on surface of the biosorbent

= Surface bound multi-metals Ni(II), Co(II), Zn(II), Cd(II)

= Free binding sites of the biosorbent: where physical adsorption of glyphosate could be occurred

(i)   = Formation of monodentate surface complexation with PO₄ group of glyphosate and metals

(ii)   = Formation of binuclear or bidentate surface complexation

(iii)   = Formation of dense-packed mononuclear surface complexation at increased glyphosate concentration

FIGURE 4.3 Schematic representation of glyphosate adsorptive mechanisms onto multi-metal laden biosorbent (Ramrakhiani et al., 2019).

4.5 REMOVAL OF AGROCHEMICALS VIA DEGRADATION

Degradation means the breaking of larger molecules into smaller ones. In natural conditions, the organic agrochemicals can degrade very slowly through processes including hydrolysis, photolysis, and biodegradation. These processes are catalyzed to achieve large-scale contaminants degradation for water and soil treatment (Zhang et al., 2022). Recently, a wide range of low-cost catalysts derived from the materials of bio-origin have been extensively utilized for agrochemicals degradation as well

(Zhang et al., 2022). For example, biochar has been extensively used in hydrolysis, photolysis, biodegradation, and catalytic oxidations for agrochemical degradation (Zhang et al., 2022). The biochar properties such as aromaticity and high functionality endow it with high electron transport capacity (Zhang et al., 2022).

4.5.1 BIODEGRADATION

Biodegradation is one of the most important processes for contaminant degradation in the environment. Biodegradation is also one of the most versatile, highly adaptable environment-friendly process with wide applications. Enzyme-immobilized materials have shown versatility in contaminants degradation, including agrochemicals (Yañez-Ocampo et al., 2009, Chen et al., 2019). For example, bacterial consortium immobilized on alginate beads was successfully utilized for the degradation of methyl parathion and tetrachlorvinphos (Yañez-Ocampo et al., 2009). Immobilizing the bacterial consortium in alginate beads also enhanced its viability to 11 days as compared to 6 days in suspension (Yañez-Ocampo et al., 2009). Similarly, laccase-immobilized peanut shell and wheat straw biosorbents were used for the treatment of nine different pesticides (Chen et al., 2019).

4.5.2 HYDROLYSIS

Hydrolysis is an important organic compound degradation pathway in aqueous systems. Modern pesticides are designed to be easily hydrolysable functional groups, but high mineralization is not possible in natural systems. Carbonaceous materials of bio-origin such as biochar can promote pesticide hydrolysis by increasing the pH and dissolved metal ion concentration and by providing heterogenous surface sites leading to alkali catalysis, homogenous dissolved metal ion catalysis, and heterogenous catalysis (Ding et al., 2017, 2020). Pesticides (carbaryl and atrazine) hydrolysis was catalyzed in biochar suspension and was first documented in 2013 (Zhang et al., 2013). The authors summarized the role of alkali and metal ions (behaving as Lewis acids) in the catalysis of atrazine and carbaryl hydrolysis generated by biochar in the suspension (Zhang et al., 2013). The catalytic capability of the biochar is proportional to its ash content, which is affected by the precursor biomass and pyrolysis conditions (Zhang et al., 2013). Another study of the hydrolysis of neonicotinoid insecticides also proceeds through alkali and metal ion hydrolysis (Zhang et al., 2018).

4.5.3 PHOTOLYSIS/PHOTODEGRADATION

Photolysis is an important natural process, in which molecules degrade directly by absorbing the light or indirectly in the presence of photosensitizers. Biomass-derived carbonaceous materials such as biochar, and hydrochar can effectively absorb light energy and contaminants. This promotes the photodegradation of contaminants by bringing the contaminants and active sites (such as OH and O2-•) closer (Zhang et al., 2022). Both biochar and biochar-derived dissolved organic matter (DOM) in aqueous systems can facilitate photodegradation by producing a high number of active species that provide effective electron transport (Serelis et al., 2021). Previous studies have also demonstrated the effectiveness of biomass-derived carbonaceous materials in pesticide photodegradation (Serelis et al., 2021, Zhang et al., 2022, Wang et al., 2019). Enhanced methylparaben degradation (97.4%) was achieved in the presence of biochar under natural sunlight (Kumar et al., 2017). The authors proposed the role of reactive oxygen species generated by the biochar carbon matrices and DOM upon irradiation of solar/ultraviolet irradiation (Kumar et al., 2017). Similarly, in another study, imidacloprid photodegradation was enhanced significantly in the presence of biochar-derived DOM due to the generation of $^1O_2$ (Zhang et al., 2020).

Metal-doped biochars can have enhanced pesticide photodegradation (Samy et al., 2022). ZnO/biochar composite catalyst with photodegradable capacity was developed using sludge and applied for malathion, lambda-cyhalothrin, and oxamyl photodegradation (Samy et al., 2022). The catalyst showed an increase in malathion, lambda-cyhalothrin, and oxamyl removal efficiencies from 30%, 38%, and 24% for pristine ZnO to 70%, 55%, and 46%, respectively, for ZnO/biochar composite (Samy et al., 2022).

4.5.4 CATALYTIC DEGRADATION

The biochar has high aromaticity which provides π-electron systems and abundant oxygen-containing functionalities, including carbonyl, hydroxyl, phenol, hydroquinone, quinone, lactone, and esters (Wang et al., 2019, Patel et al., 2019, 2021, 2022b). Biochar also has persistent free radicals formed during pyrolysis. This makes biochar an excellent electron donor and acceptor, which suggests its potential for contaminants degradation, including pesticides (Wang et al., 2019, Oh et al., 2013). Thus, biochar can catalyze oxidants including dissolved oxygen, ozone, hydrogen peroxide, and persulfate to degrade organic contaminants, including agrochemicals (Wang et al., 2019). In a fast-catalyzed metolachlor degradation study, nitrogen components and surface ketones (-C=O) groups on the surface of biochar act as active sites for HSO_5^- decomposition for enhanced degradation (Ding et al., 2020). In other studies, aromatic structures conjugated π-electrons and phenolic -OH groups on the surface of biochar catalyze the direct reduction of nitro herbicides (Fang et al., 2014, 2015).

4.6 AGROCHEMICAL REMEDIATION IN SOIL

Materials of bio-origin were also utilized for the organic and inorganic contaminants immobilization in soil successfully (Khalid et al., 2020). Carbonaceous materials such as biochar and hydrochar have emerged as a promising solution to pesticide-contaminated soil management (Cara et al., 2022). Biochar can strongly adsorb agrochemicals in soil, thus reducing their bioavailability (Zhelezova et al., 2017, Kumari et al., 2016, Ren et al., 2016). Other materials such as biomass and peat can also be utilized as biomixtures to immobilize pesticides in soil and water (Levio-Raiman et al., 2021). In a study, sodium alginate-based composite microspheres were synthesized for the controlled release of pesticides in agricultural soils (Du et al., 2023).

Utilizing materials of bio-origin in soil can significantly reduce the bioavailability of pesticides and herbicides via adsorption and different degradation processes (Khalid et al., 2020, Chen et al., 2019). For example, biochar in soil promotes the strong adsorption and enhanced mineralization of atrazine through microbial degradation (Jablonowski et al., 2013a, b). The studies suggest that biochar adsorbed pesticide (^{14}C-labeled atrazine) is degraded through the involvement of multiple mechanisms simultaneously. Biochar provides organic carbon, P and N in soil, and the shelter for microbes, which enhances microbial growth and reproduction, resulting in enhanced microbial activity. This enhanced atrazine microbial degradation (Jablonowski et al., 2013a, b). Similar results were also obtained in another study, where biochar prepared at 300°C promoted adsorption as well as microbial degradation of thiacloprid by increasing N and C sources for microbes (Zhang et al., 2018). The study also suggested that the role of oxygen-containing functional groups of biochar encourage electron transport in microbially mediated redox reactions responsible for pesticide degradation (Zhang et al., 2018). Similar roles of oxygen-containing functional groups of biochar in pentachlorophenol degradation were also reported (Tong et al., 2014). This led to agrochemical remediation by adsorption coupled with mineralization with biochar application.

In a study, nine pesticide residues were adsorbed and degraded in water and soil using biosorbent immobilized laccase (Chen et al., 2019). Peanut shell and wheat straw biosorbents were used as support to immobilize laccase and degrade pesticides (atrazine, isoproturon, bensulfuron-methyl, mefenacet, prometryn, penoxsulam, prochloraz, nitenpyram, and pyrazosulfuron-ethyl) in soil with redox mediator syringaldehyde (Chen et al., 2019). A biosorbent dose of 50 g/kg soil for peanut shell immobilized laccase and wheat straw immobilized laccase showed pesticide degradation rates of 20.9%–92.9% and 14.7%–92.0%, respectively, in soil (Chen et al., 2019). These studies showed that materials of bio-origin have enormous potential for the remediation of hazardous agrochemicals from soil.

4.7 CONCLUSIONS AND FUTURE PERSPECTIVES

Agrochemicals, especially pesticides, herbicides, and other biocides, in aquatic and terrestrial systems pose a serious threat to the environment and living beings. Low-cost adsorbents derived from the materials of bio-origin can be a sustainable solution for the problem of agrochemical contamination in environmental matrices. This chapter summarizes the development and application of different materials of bio-origin for agrochemical remediations. Materials of bio-origin are derived from the biomass of microbes, plants, and animals. These materials have certain advantages of low cost, natural abundance, ease of access, biodegradability, and ease of modification into materials with enhanced properties. Alginate, chitin, chitosan, peat, biomass, biochar, and cellulose are some of the commonly used materials of bio-origin. These materials have shown efficient removal of different agrochemicals through adsorption in aquatic and soil systems. Furthermore, these materials can also catalyze and play a significant role in different degradation processes for agrochemical degradation. Thus, utilizing materials of bio-origin would provide enormous benefits, including solid waste management and environmental remediations.

REFERENCES

Abdelhameed, R. M., Abdel-Gawad, H. & Emam, H. E. 2021. Macroporous Cu-Mof@ Cellulose Acetate Membrane Serviceable in Selective Removal of Dimethoate Pesticide from Wastewater. *Journal of Environmental Chemical Engineering*, 9, 105121.

Abida, B., Harikrishna, S. & Irfanulla, K. 2009. A Survey of Persistant Organochlorine Pesticides Residues in Some Streams of the Cauvery River, Karnataka, India. *International Journal of Chemtech Research*, 1, 237–244.

Agostini De Moraes, M., Cocenza, D. S., Da Cruz Vasconcellos, F., Fraceto, L. F. & Beppu, M. M. 2013. Chitosan and Alginate Biopolymer Membranes for Remediation of Contaminated Water with Herbicides. *Journal of Environmental Management*, 131, 222–227.

Ahmad, M., Rajapaksha, A. U., Lim, J. E., Zhang, M., Bolan, N., Mohan, D., Vithanage, M., Lee, S. S. & Ok, Y. S. 2014. Biochar as a Sorbent for Contaminant Management in Soil and Water: A Review. *Chemosphere*, 99, 19–33.

Akhtar, M., Iqbal, S., Bhanger, M. I., Zia-Ul-Haq, M. & Moazzam, M. 2009. Sorption of Organophosphorous Pesticides onto Chickpea Husk from Aqueous Solutions. *Colloids and Surfaces B: Biointerfaces*, 69, 63–70.

Aris, N. I. F., Rahman, N. A., Wahid, M. H., Yahaya, N., Abdul Keyon, A. S. & Kamaruzaman, S. 2020. Superhydrophilic Graphene Oxide/Electrospun Cellulose Nanofibre for Efficient Adsorption of Organophosphorus Pesticides from Environmental Samples. Royal Society Open Science, 7, 192050.

Aroguz, A. Z. 2006. Kinetics and Thermodynamics of Adsorption of Azinphosmethyl from Aqueous Solution onto Pyrolyzed (At 600°C) Ocean Peat Moss (Sphagnum Sp.). *Journal of Hazardous Materials*, 135, 100–105.

Ashesh, A., Singh, S., Linthoingambi Devi, N. & Chandra Yadav, I. 2022. Organochlorine Pesticides in Multi-Environmental Matrices of India: A Comprehensive Review on Characteristics, Occurrence, and Analytical Methods. *Microchemical Journal*, 177, 107306.

Ateia, M., Helbling, D. E. & Dichtel, W. R. 2020. Best Practices for Evaluating New Materials as Adsorbents for Water Treatment. *ACS Materials Letters*, 2, 1532–1544.

Aziz, K., El Achaby, M., Mamouni, R., Saffaj, N. & Aziz, F. 2023. A Novel Hydrogel Beads Based Copper-Doped Cerastoderma Edule Shells@Alginate Biocomposite for Highly Fungicide Sorption from Aqueous Medium. *Chemosphere*, 311, 136932.

Batool, S., Shah, A. A., Abu Bakar, A. F., Maah, M. J. & Abu Bakar, N. K. 2022. Removal of Organochlorine Pesticides Using Zerovalent Iron Supported on Biochar Nanocomposite from Nephelium Lappaceum (Rambutan) Fruit Peel Waste. *Chemosphere*, 289, 133011.

Borja-Urzola, A.-D.-C., García-Gómez, R. S., Bernal-González, M. & Durán-Domínguez-De-Bazúa, M.-D.-C. 2021. Chitosan-Calcite from Shrimp Residues: A Low-Cost Adsorbent for Three Triazines Removal from Aqueous Media. *Materials Today Communications*, 26, 102131.

Bouwer, H. 1989. Agriculture and Groundwater Quality. *Civil Engineering*, 59, 60.

Cao, Y., Jiang, S., Zhang, Y., Xu, J., Qiu, L. & Wang, L. 2021. Investigation into Adsorption Characteristics and Mechanism of Atrazine on Nano-Mgo Modified Fallen Leaf Biochar. *Journal of Environmental Chemical Engineering*, 9, 105727.

Cara, I. G., Ţopa, D., Puiu, I. & Jităreanu, G. 2022. Biochar a Promising Strategy for Pesticide-Contaminated Soils. Agriculture, 12, 1579.

Chen, X., Zhou, Q., Liu, F., Peng, Q. & Teng, P. 2019. Removal of Nine Pesticide Residues from Water and Soil by Biosorption Coupled with Degradation on Biosorbent Immobilized Laccase. *Chemosphere*, 233, 49–56.

Choudhary, V., Patel, M., Pittman, C. U. & Mohan, D. 2020. Batch and Continuous Fixed-Bed Lead Removal Using Himalayan Pine Needle Biochar: Isotherm and Kinetic Studies. *ACS Omega*, 5, 16366–16378.

Clarke, B. O. & Smith, S. R. 2011. Review of 'Emerging' Organic Contaminants in Biosolids and Assessment of International Research Priorities for the Agricultural Use of Biosolids. *Environment International*, 37, 226–247.

Covert, S. A., Shoda, M. E., Stackpoole, S. M. & Stone, W. W. 2020. Pesticide Mixtures Show Potential Toxicity to Aquatic Life in U.S. Streams, Water Years 2013–2017. *Science of the Total Environment*, 745, 141285.

De Aguiar, T. R., Guimarães Neto, J. O. A., Şen, U. & Pereira, H. 2019. Study of Two Cork Species as Natural Biosorbents for Five Selected Pesticides in Water. Heliyon, 5, E01189.

Ding, D., Yang, S., Qian, X., Chen, L. & Cai, T. 2020. Nitrogen-Doping Positively Whilst Sulfur-Doping Negatively Affect the Catalytic Activity of Biochar for the Degradation of Organic Contaminant. Applied Catalysis B: Environmental, 263, 118348.

Ding, Y., Liu, Y., Liu, S., Huang, X., Li, Z., Tan, X., Zeng, G. & Zhou, L. 2017. Potential Benefits of Biochar in Agricultural Soils: A Review. *Pedosphere*, 27, 645–661.

Du, Y., Zhang, Q., Yu, M., Jiao, B., Chen, F. & Yin, M. 2023. Sodium Alginate-Based Composite Microspheres for Controlled Release of Pesticides and Reduction of Adverse Effects of Copper in Agricultural Soils. Chemosphere, 313, 137539.

Essandoh, M., Wolgemuth, D., Pittman, C. U., Mohan, D. & Mlsna, T. 2017a. Adsorption of Metribuzin from Aqueous Solution Using Magnetic and Nonmagnetic Sustainable Low-Cost Biochar Adsorbents. *Environmental Science and Pollution Research*, 24, 4577–4590.

Essandoh, M., Wolgemuth, D., Pittman, C. U., Mohan, D. & Mlsna, T. 2017b. Phenoxy Herbicide Removal from Aqueous Solutions Using Fast Pyrolysis Switchgrass Biochar. *Chemosphere*, 174, 49–57.

Fang, G., Gao, J., Liu, C., Dionysiou, D. D., Wang, Y. & Zhou, D. 2014. Key Role of Persistent Free Radicals in Hydrogen Peroxide Activation by Biochar: Implications to Organic Contaminant Degradation. *Environmental Science & Technology*, 48, 1902–1910.

Fang, G., Liu, C., Gao, J., Dionysiou, D. D. & Zhou, D. 2015. Manipulation of Persistent Free Radicals in Biochar to Activate Persulfate for Contaminant Degradation. *Environmental Science & Technology*, 49, 5645–5653.

FAO. 2017. *World Fertilizer Trends and Outlook to 2020*. Rome: Food and Agriculture Organization of the United Nations.

Gámiz, B., Hall, K., Spokas, K. A. & Cox, L. 2019. Understanding Activation Effects on Low-Temperature Biochar for Optimization of Herbicide Sorption. *Agronomy*, 9, 588.

Gautam, R. K., Goswami, M., Mishra, R. K., Chaturvedi, P., Awashthi, M. K., Singh, R. S., Giri, B. S. & Pandey, A. 2021. Biochar for Remediation of Agrochemicals and Synthetic Organic Dyes from Environmental Samples: A Review. *Chemosphere*, 272, 129917.

Guo, Y., Yu, J., Li, X., Guo, L., Xiao, C., Chi, R., Hou, H. & Feng, G. 2022. Selective Recovery of Glyphosine from Glyphosate Mother Liquor Using a Modified Biosorbent: Competitive Substitution Adsorption. *Environmental Research*, 215, 114394.

Gupta, K., Kumar, V., Tikoo, K. B., Kaushik, A. & Singhal, S. 2020. Encrustation of Cadmium Sulfide Nanoparticles into the Matrix of Biomass Derived Silanized Cellulose Nanofibers for Adsorptive Detoxification of Pesticide and Textile Waste. *Chemical Engineering Journal*, 385, 123700.

He, Y., Zhang, B., Wu, Y., Ouyang, J., Huang, M., Lu, S., Sun, H. & Zhang, T. 2021. A Pilot Nationwide Baseline Survey on the Concentrations of Neonicotinoid Insecticides in Tap Water from China: Implication for Human Exposure. *Environmental Pollution*, 291, 118117.

Hokkanen, S., Bhatnagar, A. & Sillanpää, M. 2016. A Review on Modification Methods to Cellulose-Based Adsorbents to Improve Adsorption Capacity. *Water Research*, 91, 156–173.

Jablonowski, N. D., Borchard, N., Zajkoska, P., Fernández-Bayo, J. D., Martinazzo, R., Berns, A. E. & Burauel, P. 2013a. Biochar-Mediated [14c]Atrazine Mineralization in Atrazine-Adapted Soils from Belgium and Brazil. *Journal of Agricultural and Food Chemistry*, 61, 512–516.

Jablonowski, N. D., Krutz, J. L., Martinazzo, R., Zajkoska, P., Hamacher, G., Borchard, N. & Burauel, P. 2013b. Transfer of Atrazine Degradation Capability to Mineralize Aged 14c-Labeled Atrazine Residues in Soils. *Journal of Agricultural and Food Chemistry*, 61, 6161–6166.

Jacob, M. M., Ponnuchamy, M., Kapoor, A. & Sivaraman, P. 2022. Adsorptive Decontamination Oof Organophosphate Pesticide Chlorpyrifos from Aqueous Systems Using Bagasse-Derived Biochar Alginate Beads: Thermodynamic, Equilibrium, and Kinetic Studies. *Chemical Engineering Research and Design*, 186, 241–251.

Jain, K., Rani, P., Patel, M., Dhaka, S., Ahalawat, S., Rana, A., Mohan, D., Singh, K. P. & Dhaka, R. K. 2022. Agricultural Residue-Derived Sustainable Nanoadsorbents for Wastewater Treatment. In: Madhav, S., Singh, P., Mishra, V., Ahmed, S. & Mishra, P. K. (Eds.), *Recent Trends in Wastewater Treatment* (pp. 235–259). Cham: Springer International Publishing.

Jani, J., Raiyani, C., Mistry, J., Patel, J., Desai, N. & Kashyap, S. 1991. Residues of Organochlorine Pesticides and Polycyclic Aromatic Hydrocarbons in Drinking Water of Ahmedabad City, India. *Bulletin of Environmental Contamination and Toxicology*, 47, 381–385.

Kaushik, C., Sharma, H., Jain, S., Dawra, J. & Kaushik, A. 2008. Pesticide Residues in River Yamuna and Its Canals in Haryana and Delhi, India. *Environmental Monitoring and Assessment*, 144, 329–340.

Kaushik, C., Sharma, H. & Kaushik, A. 2012. Organochlorine Pesticide Residues in Drinking Water in the Rural Areas of Haryana, India. *Environmental Monitoring and Assessment*, 184, 103–112.

Khalid, S., Shahid, M., Murtaza, B., Bibi, I., Natasha, Asif Naeem, M. & Niazi, N. K. 2020. A Critical Review of Different Factors Governing the Fate of Pesticides in Soil under Biochar Application. *Science of the Total Environment*, 711, 134645.

Krasucka, P., Pan, B., Sik Ok, Y., Mohan, D., Sarkar, B. & Oleszczuk, P. 2021. Engineered Biochar – A Sustainable Solution for the Removal of Antibiotics from Water. *Chemical Engineering Journal*, 405, 126926.

Kumar, A., Shalini, Sharma, G., Naushad, M., Kumar, A., Kalia, S., Guo, C. & Mola, G. T. 2017. Facile Hetero-Assembly of Superparamagnetic Fe3O4/BiVO4 Stacked on Biochar for Solar Photo-Degradation of Methyl Paraben and Pesticide Removal from Soil. *Journal of Photochemistry and Photobiology A: Chemistry*, 337, 118–131.

Kumarasamy, P., Govindaraj, S., Vignesh, S., Rajendran, R. B. & James, R. A. 2012. Anthropogenic Nexus on Organochlorine Pesticide Pollution: A Case Study with Tamiraparani River Basin, South India. *Environmental Monitoring and Assessment*, 184, 3861–3873.

Kumari, K. G. I. D., Moldrup, P., Paradelo, M., Elsgaard, L. & De Jonge, L. W. 2016. Soil Properties Control Glyphosate Sorption in Soils Amended with Birch Wood Biochar. *Water, Air, & Soil Pollution*, 227, 174.

Lari, S. Z., Khan, N. A., Gandhi, K. N., Meshram, T. S. & Thacker, N. P. 2014. Comparison of Pesticide Residues in Surface Water and Ground Water of Agriculture Intensive Areas. *Journal of Environmental Health Science and Engineering*, 12, 1–7.

Lee, Y.-G., Shin, J., Kwak, J., Kim, S., Son, C., Cho, K. H. & Chon, K. 2021. Effects of Naoh Activation on Adsorptive Removal of Herbicides by Biochars Prepared from Ground Coffee Residues. *Energies*, 14, 1297.

Levio-Raiman, M., Schalchli, H., Briceño, G., Bornhardt, C., Tortella, G., Rubilar, O. & Cristina Diez, M. 2021. Performance of an Optimized Fixed-Bed Column Packed with an Organic Biomixture to Remove Atrazine from Aqueous Solution. *Environmental Technology & Innovation*, 21, 101263.

Malik, A., Ojha, P. & Singh, K. P. 2009. Levels and Distribution of Persistent Organochlorine Pesticide Residues in Water and Sediments of Gomti River (India)—A Tributary of the Ganges River. *Environmental Monitoring and Assessment*, 148, 421–435.

Mayakaduwa, S. S., Herath, I., Ok, Y. S., Mohan, D. & Vithanage, M. 2017. Insights into Aqueous Carbofuran Removal by Modified and Non-Modified Rice Husk Biochars. *Environmental Science and Pollution Research*, 24, 22755–22763.

Md Meftaul, I., Venkateswarlu, K., Dharmarajan, R., Annamalai, P. & Megharaj, M. 2020. Pesticides in the Urban Environment: A Potential Threat That Knocks at the Door. *Science of the Total Environment*, 711, 134612.

Mohan, D., Sarswat, A., Ok, Y. S. & Pittman, C. U. 2014. Organic and Inorganic Contaminants Removal from Water with Biochar, a Renewable, Low Cost and Sustainable Adsorbent – A Critical Review. *Bioresource Technology*, 160, 191–202.

Moradi Dehaghi, S., Rahmanifar, B., Moradi, A. M. & Azar, P. A. 2014. Removal of Permethrin Pesticide from Water by Chitosan–Zinc Oxide Nanoparticles Composite as an Adsorbent. *Journal of Saudi Chemical Society*, 18, 348–355.

Mutiyar, P. & Mittal, A. 2013. Status of Organochlorine Pesticides in Ganga River Basin: Anthropogenic or Glacial? *Drinking Water Engineering and Science*, 6, 69–80.

Narayanan, N., Gupta, S. & Gajbhiye, V. T. 2020. Decontamination of Pesticide Industrial Effluent by Adsorption–Coagulation–Flocculation Process Using Biopolymer-Nanoorganoclay Composite. *International Journal of Environmental Science and Technology*, 17, 4775–4786.

Ng, E.-L., Huerta Lwanga, E., Eldridge, S. M., Johnston, P., Hu, H.-W., Geissen, V. & Chen, D. 2018. An Overview of Microplastic and Nanoplastic Pollution in Agroecosystems. *Science of the Total Environment*, 627, 1377–1388.

Oh, S.-Y., Son, J.-G. & Chiu, P. C. 2013. Biochar-Mediated Reductive Transformation of Nitro Herbicides and Explosives. *Environmental Toxicology and Chemistry*, 32, 501–508.

Okoro, H. K., Pandey, S., Ogunkunle, C. O., Ngila, C. J., Zvinowanda, C., Jimoh, I., Lawal, I. A., Orosun, M. M. & Adeniyi, A. G. 2022. Nanomaterial-Based Biosorbents: Adsorbent for Efficient Removal of Selected Organic Pollutants from Industrial Wastewater. *Emerging Contaminants*, 8, 46–58.

Patel, M., Chaubey, A. K., Navarathna, C., Mlsna, T. E., Pittman, C. U. & Mohan, D. 2022a. Sorptive Removal of Pharmaceuticals Using Sustainable Biochars. In: Mohan, D., Pittman, C. U. & Mlsna, T. E. (Eds.), Sustainable Biochar for Water and Wastewater Treatment (pp. 395–427). Amsterdam: Elsevier.

Patel, M., Chaubey, A. K., Pittman, C. U. & Mohan, D. 2022b. Aqueous Ibuprofen Sorption by Using Activated Walnut Shell Biochar: Process Optimization and Cost Estimation. *Environmental Science: Advances*, 1, 530–545.

Patel, M., Kumar, R., Kishor, K., Mlsna, T., Pittman, C. U. & Mohan, D. 2019. Pharmaceuticals of Emerging Concern in Aquatic Systems: Chemistry, Occurrence, Effects, and Removal Methods. *Chemical Reviews*, 119, 3510–3673.

Patel, M., Kumar, R., Pittman, C. U. & Mohan, D. 2021. Ciprofloxacin and Acetaminophen Sorption onto Banana Peel Biochars: Environmental and Process Parameter Influences. *Environmental Research*, 201, 111218.

Pratap, T., Chaubey, A. K., Patel, M., Mlsna, T. E., Pittman, C. U. & Mohan, D. 2022. 20-Nanobiochar for Aqueous Contaminant Removal. In: Mohan, D., Pittman, C. U. & Mlsna, T. E. (Eds.), Sustainable Biochar for Water and Wastewater Treatment (pp. 667–704). Amsterdam: Elsevier.

Pratap, T., Patel, M., Pittman, C. U., Nguyen, T. A. & Mohan, D. 2021. Chapter 23: Nanobiochar: A Sustainable Solution for Agricultural and Environmental Applications. In: Amrane, A., Mohan, D., Nguyen, T. A., Assadi, A. A. & Yasin, G. (Eds.), *Nanomaterials for Soil Remediation* (pp. 501–519). Amsterdam: Elsevier.

Qamar, S. A., Qamar, M., Basharat, A., Bilal, M., Cheng, H. & Iqbal, H. M. N. 2022. Alginate-Based Nano-Adsorbent Materials – Bioinspired Solution to Mitigate Hazardous Environmental Pollutants. *Chemosphere*, 288, 132618.

Ramrakhiani, L., Ghosh, S., Mandal, A. K. & Majumdar, S. 2019. Utilization of Multi-Metal Laden Spent Biosorbent for Removal of Glyphosate Herbicide from Aqueous Solution and its Mechanism Elucidation. *Chemical Engineering Journal*, 361, 1063–1077.

Rana, A. K., Mishra, Y. K., Gupta, V. K. & Thakur, V. K. 2021. Sustainable Materials in the Removal of Pesticides from Contaminated Water: Perspective on Macro to Nanoscale Cellulose. *Science of the Total Environment*, 797, 149129.

Ranjbar Bandforuzi, S. & Hadjmohammadi, M. R. 2019. Modified Magnetic Chitosan Nanoparticles Based on Mixed Hemimicelle of Sodium Dodecyl Sulfate for Enhanced Removal and Trace Determination of Three Organophosphorus Pesticides from Natural Waters. *Analytica Chimica Acta*, 1078, 90–100.

Ray, P. 1989. *Measurements on Ganga River Water Quality-Heavy Metals and Pesticides*. Final Report. Lucknow: Industrial Toxicological Research Centre.

Ren, X., Zhang, P., Zhao, L. & Sun, H. 2016. Sorption and Degradation of Carbaryl in Soils Amended with Biochars: Influence of Biochar Type and Content. *Environmental Science and Pollution Research*, 23, 2724–2734.

Samy, M., Gar Alalm, M., Ezeldean, E., El-Dissouky, A., Badr, N. B. E., Al-Muhtaseb, A. A., Alhajeri, N. S., Osman, A. I. & Tawfik, A. 2022. Solar-Light-Driven Zno/Biochar Treatment of Pesticides Contaminated Wastewater: A Practical and Computational Study. Energy Science & Engineering, 10, 4708–4725.

Schwantes, D., Gonçalves Jr, A. C., Fuentealba, D., Hornos Carneiro, M. F., Tarley, C. R. T. & Prete, M. C. 2022. Removal of Chlorpyrifos from Water Using Biosorbents Derived from Cassava Peel, Crambe Meal, and Pinus Bark. *Chemical Engineering Research and Design*, 188, 142–165.

Serelis, K., Mantzos, N., Meintani, D. & Konstantinou, I. 2021. The Effect of Biochar, Hydrochar Particles and Dissolved Organic Matter on the Photodegradation of Metribuzin Herbicide in Aquatic Media. *Journal of Environmental Chemical Engineering*, 9, 105027.

Shah, J., Jan, M. R. & Tasmia 2018. Magnetic Chitosan Graphene Oxide Composite for Solid Phase Extraction of Phenylurea Herbicides. *Carbohydrate Polymers*, 199, 461–472.

Sharma, A., Shukla, A., Attri, K., Kumar, M., Kumar, P., Suttee, A., Singh, G., Barnwal, R. P. & Singla, N. 2020. Global Trends in Pesticides: A Looming Threat and Viable Alternatives. *Ecotoxicology and Environmental Safety*, 201, 110812.

Singh, S., Kumar, V., Datta, S., Dhanjal, D. S., Sharma, K., Samuel, J. & Singh, J. 2020. Current Advancement and Future Prospect of Biosorbents for Bioremediation. *Science of the Total Environment*, 709, 135895.

Suhas, Gupta, V. K., Carrott, P. J. M., Singh, R., Chaudhary, M. & Kushwaha, S. 2016. Cellulose: A Review as Natural, Modified and Activated Carbon Adsorbent. *Bioresource Technology*, 216, 1066–1076.

Sutirman, Z. A., Sanagi, M. M. & Wan Aini, W. I. 2021. Alginate-Based Adsorbents for Removal of Metal Ions and Radionuclides from Aqueous Solutions: A Review. *International Journal of Biological Macromolecules*, 174, 216–228.

Tong, H., Hu, M., Li, F. B., Liu, C. S. & Chen, M. J. 2014. Biochar Enhances the Microbial and Chemical Transformation of Pentachlorophenol in Paddy Soil. *Soil Biology and Biochemistry*, 70, 142–150.

Vimal, V., Patel, M. & Mohan, D. 2019. Aqueous Carbofuran Removal Using Slow Pyrolyzed Sugarcane Bagasse Biochar: Equilibrium and Fixed-Bed Studies. *Rsc Advances*, 9, 26338–26350.

Vithanage, M., Mayakaduwa, S. S., Herath, I., Ok, Y. S. & Mohan, D. 2016. Kinetics, Thermodynamics and Mechanistic Studies of Carbofuran Removal Using Biochars from Tea Waste and Rice Husks. *Chemosphere*, 150, 781–789.

Wang, R.-Z., Huang, D.-L., Liu, Y.-G., Zhang, C., Lai, C., Wang, X., Zeng, G.-M., Gong, X.-M., Duan, A., Zhang, Q. & Xu, P. 2019. Recent Advances in Biochar-Based Catalysts: Properties, Applications and Mechanisms for Pollution Remediation. *Chemical Engineering Journal*, 371, 380–403.

Wang, X., Hou, X., Zou, P., Huang, A., Zhang, M. & Ma, L. 2022. Cationic Starch Modified Bentonite-Alginate Nanocomposites for Highly Controlled Diffusion Release of Pesticides. *International Journal of Biological Macromolecules*, 213, 123–133.

Wiśniewska, M. & Nowicki, P. 2020. Peat-Based Activated Carbons as Adsorbents for Simultaneous Separation of Organic Molecules from Mixed Solution of Poly(Acrylic Acid) Polymer and Sodium Dodecyl Sulfate Surfactant. *Colloids and Surfaces A: Physicochemical and Engineering Aspects*, 585, 124179.

Yadav, I. C., Devi, N. L., Syed, J. H., Cheng, Z., Li, J., Zhang, G. & Jones, K. C. 2015. Current Status of Persistent Organic Pesticides Residues in Air, Water, and Soil, and Their Possible Effect on Neighboring Countries: A Comprehensive Review of India. *Science of the Total Environment*, 511, 123–137.

Yañez-Ocampo, G., Sanchez-Salinas, E., Jimenez-Tobon, G. A., Penninckx, M. & Ortiz-Hernández, M. L. 2009. Removal of Two Organophosphate Pesticides by a Bacterial Consortium Immobilized in Alginate or Tezontle. *Journal of Hazardous Materials*, 168, 1554–1561.

Yu, J., Wang, J., & Jiang, Y. (2017). Removal of uranium from aqueous solution by alginate beads. *Nuclear Engineering and Technology*, 49(3), 534–540.

Zhang, P., Huang, P., Ma, M., Meng, X., Hao, Y. & Sun, H. 2022. Chapter 11: Effects of Biochar on the Environmental Behavior of Pesticides. In: Tsang, D. C. W. & Ok, Y. S. (Eds.), *Biochar in Agriculture for Achieving Sustainable Development Goals* (pp. 129–135). Amsterdam: Academic Press.

Zhang, P., Shao, Y., Xu, X., Huang, P. & Sun, H. 2020. Phototransformation of Biochar-Derived Dissolved Organic Matter and the Effects on Photodegradation of Imidacloprid in Aqueous Solution Under Ultraviolet Light. *Science of the Total Environment*, 724, 137913.

Zhang, P., Sun, H., Min, L. & Ren, C. 2018. Biochars Change the Sorption and Degradation of Thiacloprid in Soil: Insights into Chemical and Biological Mechanisms. *Environmental Pollution*, 236, 158–167.

Zhang, P., Sun, H., Yu, L. & Sun, T. 2013. Adsorption and Catalytic Hydrolysis of Carbaryl and Atrazine on Pig Manure-Derived Biochars: Impact of Structural Properties of Biochars. *Journal of Hazardous Materials*, 244–245, 217–224.

Zhelezova, A., Cederlund, H. & Stenström, J. 2017. Effect of Biochar Amendment and Ageing on Adsorption and Degradation of Two Herbicides. *Water, Air, & Soil Pollution*, 228, 216.

Zhou, Q., Wang, W., Liu, F. & Chen, R. 2022. Removal of Difenoconazole and Nitenpyram by Composite Calcium Alginate Beads during Apple Juice Clarification. *Chemosphere*, 286, 131813.

5 Biochemical and Molecular Aspects of Phytoremediation toward Mitigation of Heavy Metals

Dolly Kain and Atul Arya

5.1 INTRODUCTION

Phytoremediation comes under in situ rectification technology that is based on the inherent capabilities of plants. It is constructed on the phenomenon of clean nature by using nature and can be defined as ecologically friendly and solar energy-driven technology (UNEP, undated). Plants can be used to clean different contaminants like heavy metals, pesticides, explosives, and oil spills. They are also helpful in preventing the spread of contaminants from one place to the other because of their root system. This technology is very effective in the tropical regions due to prevailing meteorological circumstances, which increase the growth of plants and also enhance the activity of microbes (Zhang et al., 2010). It involves the usage of metabolic and absorption properties and also transport systems of plants and results in the sequestration and degradation of the contaminants. Conventional treatments such as soil cleansing, combustion, excavation, and disposal are employed to rehabilitate areas of contamination, but these procedures are extremely expensive (Pilon-smith, 2009). But phytoremediation is a less expensive technology, easy to implement, ecologically friendly, and hence the most favorite technique of remediation nowadays. It comprises two components that include root microbes and plants. Plants are capable of remediating a wide range of organic, xenobiotic, synthetic, hydrocarbon, pesticide, and heavy metal contaminants (Suresh and Ravi Shankar, 2004). According to the Environmental Action Group, the Chinese town of Linfen serves as an example of contamination and is representative of the pollution that many Chinese cities face. In Ranipet (India), 3.5 million people are affected by tannery waste, including azo dyes and hexavalent chromium. Precious metals and coal mining form the mainstays of the economy of numerous nations. Developing countries such as Brazil, India, China, and Peru accord to the world's largest mining industry. Mining activities affect human health and the environment.

Elements with masses that range between 63.54 and 200.59 are classified as heavy metals. They are major pollutants of water bodies and soil. They are imperishable

DOI: 10.1201/9781003407317-5

compounds and endure for a considerable amount of time in the surroundings. Heavy metals can be of two types—beneficial (micronutrient) and toxic heavy metals. Micronutrients such as Fe, Co, Zn, and Cu are needed in minute amounts for metabolism, but in high concentrations, they become toxic. Examples of heavy metals that are not necessary include arsenic, mercury, lead, chromium, and cadmium. The two primary sources of contamination with heavy metals are geological processes and the actions of humans (Dembitsky, 2003). Human-related activities include industrial waste, manufacturing of fuel, mining, metallurgical processes, armed forces operations, chemicals for agriculture, small-scale businesses (such as manufacturing batteries and cable coatings), brick-making facilities, and coal combustion (Zhen-Guo etal., 2002). Another major contributor is municipal waste, which is abandoned on roadways or utilized as landfills, and wastewater is also used as a source of irrigation. Waste products provide nutrients, but they additionally contain carcinogens and harmful metals. Additional variables include the excessive use of fertilizers, fungicides, and pesticides (Zhen-Guo et al., 2002). Heavy metals cause drastic negative effects in plants that include (i) disturbance of protein structure by forming bonds with sulfhydryl and functional groups of important molecules, (ii) disruption of the functionality of essential elements in biomolecules such as pigments or enzymes, (iii) generation of cytotoxic compounds like methylglyoxal causes oxidative stress leading to plant cell death (iv) increased formation of reactive oxygen species (ROS), i.e., superoxide free radicals (O_2^-) or non-free radical species like singlet oxygen (O_2) and H_2O_2, which are highly toxic to plants, (v) reduction of the chlorophyll content of plants by damaging the light-harvesting system and photosystem-II of photosynthesis (Fernandes and Henriques, 1991).

5.1.1 HYPERACCUMULATORS

Plants may show different growth strategies in heavy metal-affected soils and can therefore be organized into three types: metal excretors, metal scavengers, and metal hyperaccumulators (Baker and Walker, 1989). Excluders are defined as plants that contain heavy metals in their roots but whose shoot/root ratio is consistently lower than 1. The roots contain large amounts of iron but prevent heavy metals from entering the body. Example: *Willow* species. Metal detectors store metal in the tissues above, similar to the surrounding soil. Example: *Eucalyptus* stores gold. Hyperaccumulator plants serve as models for phytoremediation methods because they are more permissive to heavy metals than other plants. Hyperaccumulators are plants that can store 100 times more iron than ordinary non-accumulators. Ni-rich strains of iron hyperaccumulator plants were first described by Jaffre et al. (2013). Nickel hyperaccumulators are species that accumulate at least 1000 mg/kg of nickel (dry weight) in their leaves while growing in their environment. Therefore, hyperenriched species of other metals will be identified. Baker and colleagues first considered the use of hyperaccumulator bacteria in phytoremediation (phytoextraction) because these plants could accumulate high amounts of soil. It makes up about 0.2% of angiosperms. Iron overaccumulation and tolerance are genetic traits. The heavy metal hyperaccumulation mechanism pivots on the plant, conditions of soil, i.e., pH, cation exchange ability, content of organic matter, and type of heavy metal (Spinoza-Quinones et al., 2005). Some metals considered necessary for proper plant growth (such as iron and copper) can be harmful depending on their oxidation state, complex form, dosage, and proximity (Beyersmann and Hartwig, 2008). Baker et al. (1991) performed the experimentation on the phytoextraction of zinc and cadmium. Some important hyperaccumulator families include Flacourtiaceae, Fabaceae, Poaceae, Lamiaceae, Volaceae, and Euphorbiaceae. The hyperaccumulation of iron occurs in the range of 0.2% of angiosperms, especially Brassicaceae. Twenty-six species of hyperaccumulator cobalt belonging to the families Lamiaceae, Scrophulariaceae, Asteraceae, and Leguminosae have been reported (Baker et al., 2000). In 1865, excessive zinc accumulation was observed for the first time in the *Noccaea caerulescent* (same name as *Thraspi caerulescent*) species of the Brassicaceae family (Reeves and Baker, 2000). *Sedum alfrediii* is the sole cadmium hyperaccumulator, which is external to the Brassicaceae family (Deng et al., 2007). In addition to ferns (Ma et al., 2001), excessive arsenic accumulation has also been observed in two species of the Brassicaceae family (Karimi et al., 2009).

Accumulators and hyperaccumulators are plants that contain higher levels of iron than other plants. Hyperaccumulation occurs when metal ions accumulate at levels greater than 0.1%–1% of total plant dry weight. They are very adaptable to iron-rich soils. Iron impacts plant physiology by either enhancing or restricting proliferation. Some metals are needed at high concentrations to have properties or osmotic roles, while others are required at low concentrations to exert their role as cofactors for certain enzymes. Plants enhance iron tolerance via apoplast or symplast detoxifying strategies (Ma et al., 2001; Pilon-Smith et al., 2009). Co binds strongly to roots and becomes immobilized. Co's chemical properties are similar to those of nickel. Co enters cells through the plasma membrane transporter i.e. IRT-1 (Pilon-Smith et al., 2009). Sturdy adsorption by root apoplasts is the driving force of iron extraction from soil. For example, *Astragalus bisulcatus* has a selenocysteine methyltransferase specific for selenium accumulation. It binds to mercury and nitrogen-rich ligands, i.e., amino acids, blocking functional groups of enzymes and damaging the cell integrity and membranes of organelles (Ochiai, 1987). Fe and Cu can be very harmful when present in large amounts because they are involved in redox reactions and produce hydroxyl radicals that are toxic to cells (Stochs and Bagchi, 1995). Cadmium is a non-redox metal and is highly phytotoxic, causing growth inhibition and death of plant cells. It causes changes in the profile of lipids (Ouariti et al., 1997) and influences the activity of the H+ ATPase membrane (Fodor et al., 1995). It also causes harm to the energy supply, lowers chlorophyll concentration (Siedlecka and Baszynsky, 1993), and hinders stomatal regulation (Barceló and Poschenreider, 1990).

5.2 BIOCHEMICAL AND MOLECULAR ASPECTS OF PHYTOREMEDIATION

Heavy metals form many groups, and the toxicity varies depending on the metal and its concentration. Many of these elements and their soluble forms are highly toxic. Their appearance in the air, water, and soil can lead to major issues for living beings. Bioaccumulation of heavy metals in food (nickel, copper, iron, etc.) is a serious threat and thus plant metabolism evolves over time and has developed sensitive mechanisms that regulate the movement of these heavy metals in cells. Plants can accumulate nonessential metals such as cadmium. The method through which it happens is unknown, but one possibility is that Cd is adsorbed to form analogs of other divalent cations such as Zn^+. The transfer of heavy metals is the movement of ions after entering, and the transportation of iron-containing acid from the roots to the soil is called translocation. Pressure from roots and the transpiration of leaves are the two mechanisms that regulate translocation. After translocation, iron is reabsorbed into leaf cells (Beyersmann and Hartwig, 2008). Ions cannot cross the plasma membrane, so the transport of iron using proteins, which act as a mediator to enter into cells, are known as carriers. Transport proteins are transmembrane proteins that have extracellular domains, transmembrane domains, and intracellular domains. The ions bind to extracellular areas that accept specific ions. Transmembrane structures transfer ions present in the extracellular space to the intracellular hydrophobic environment. Only a fraction of the total amount of ions absorbed by the roots is absorbed by the plant. Most of the ions are absorbed by the stem cell wall (e.g., Pb accumulator). Ions bound to the cell wall cannot be transported and therefore cannot be harvested (phytoextraction) (Beyersmann and Hartwig, 2008; Fuentus et al., 2000).

5.2.1 BIOCHEMICAL ASPECTS OF PHYTOREMEDIATION

Heavy metals can cause biochemical changes in plants. These changes can be better understood from the following key steps in heavy metals (Figure 5.1) (Yang et al., 2005):

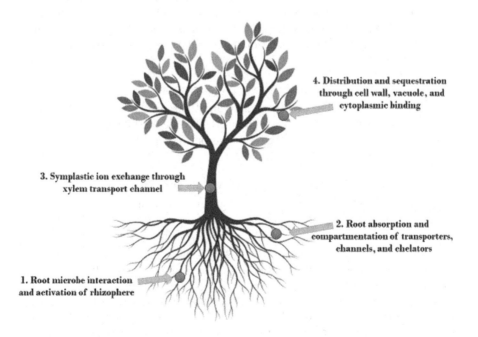

4. Distribution and sequestration through cell wall, vacuole, and cytoplasmic binding

3. Symplastic ion exchange through xylem transport channel

2. Root absorption and compartmentation of transporters, channels, and chelators

1. Root microbe interaction and activation of rhizophere

FIGURE 5.1 Major steps involved in the hyperaccumulation of heavy metals using plants.

1. **Adsorption:** Hyperaccumulators of plants have extensive roots in the area, resulting in an unusual concentration of metals and nutrients, as in Indian mustard. The adsorption process depends on two factors: (1) bases and (2) iron bioavailability. The interaction between bacteria and roots can increase the amount of iron because bacteria release some organic acids, enzymes, amino acids, and phytochelatins, causing the breakdown and adsorption of metals.
2. **Accumulation and Transport:** Iron-binding proteins and peptides produced by plants have a key function in the accumulation of iron. These proteins/peptides are specific for metal and retain only toxic metals, not essential metals. Aggregation and transport can be improved by the use of chelating agents (natural and synthetic). Metal-accumulating plants can be divided into three groups: (i) Cu/Co, (ii) Zn/Cd/Pb, and (iii) Ni (Raskin et al., 1994). It is facilitated by transport proteins that mediate transport across the plasma membrane. There is a special binding in the carrier that binds to the complement region and carries it into the cell. Some proteins involved in accumulation and transport in Arabidopsis include CPx-type heavy metal H+ ATPase, NRAMP (innate resistance-associated macrophage protein family), CDF-cation diffusion exciton, and ZIP (zinc ion permease) family.
3. **Translocation:** The process of transporting iron ions from roots to shoots is called translocation. Storage in plants depends on the retention of heavy metals in specialized tissues such as trichomes and cellular compartments such as vacuoles. Membrane transport systems play an important role in the translocation process. Plants can also secrete phytosiderophores to prevent iron ion deficiency; H+ ATPase can catalyze the transport of Zn and Cd. MATE (multidrug and toxin efflux) is the transport of small organic molecules.

Several heavy metals impede the production of biochemicals, which has been thoroughly researched and assessed (Fernandes and Henriques, 1991). Plants' pigments used for photosynthesis from various groups show varied metal tolerance (Vajpayee et al., 2001). One consequence of heavy metals on chlorophyll is a reduction in fluorescence (Atal et al., 1991; El-Sheek, 1992).

5.2.1.1 Example of Three Major Heavy Metals and Their Effects

Cadmium reduced the photosynthetic pigments as seen in two species viz. *Potamogeton crispus* and *Myriophyllum heterophyllum*. In cadmium (Cd), hyperaccumulator sp. *Atriplex halimus* subsp. *Schweinfurthii* results in chlorophyll reduction, transpiration using stomata, and hydraulic conductivity of roots (Nedjimi and Daoud, 2009). Chromium results in modifications of pigments and amino acids. Cr (VI) causes toxicity to D-aminolevulinic acid dehytdratase (involved in chlorophyll biosynthesis). Cr (VI) might swap out the magnesium in the enzyme's active site and result in the depletion of chlorophyll. Cr also inhibits chlorophyll biosynthesis (Barcelo et al., 1986). Cr can result in the degradation of carotenoids. But according to Vajpayee et al. (2001), carotenoid content was increased in chromium-treated plants. Thus, the impact of heavy metals on the amount of carotenoid was plant- and metal-specific. Mercury could affect the normal function of proteins using two binding sites of proteins. The binding sites of proteins with deformation in the chain result in precipitation (Clarkson, 1972). On the other hand, Hg impacts both the daylight and darkness methods of photosynthesis and results in the inhibition of electron transport chain (ETC), oxygen evolution, and quenching of fluorescence based on chlorophyll in photosystem-II. Hg can replace Mg as the central atom of chlorophyll which results in photosynthesis inhibition (Krupa and Baszynski, 1995).

5.2.2 Molecular Aspects of Phytoremediation

Coordinated networks of molecular mechanisms are known to improve the phytoremediation process. Some of the molecular events involved in metal uptake are explained (Figure 5.2):

1. **Heavy Metal Perceptivity and Transduction of Signal:** Protein kinases that resemble receptors mediate the reception of signals outside the cell. It has been demonstrated that Cr^{2+}, Cd^{2+}, and Cu^{2+} may trigger the barley gene that codes for lysine motif receptor-like kinase (Gleba et al., 1999).
2. **Signaling Cascade:** Stress signals induced by heavy metals involved in calcium changes the MAPK cascade and transcriptional regulation of the genes involved in stress response (Shao et al., 2008). Barley receptor-like kinases expression is regulated by Ca^{2+} level. Heavy metals were found to interfere in calmodulin pathway as they replace Ca^{2+}. Some evidence suggests that treating tobacco cells with cadmium and lupines roots using Pb^{2+} causes the formation of ROS reactive oxygen species such as H_2O_2, resulting in an oxidative burst mediated by the Ca/calmodulin system (Shao et al., 2010).

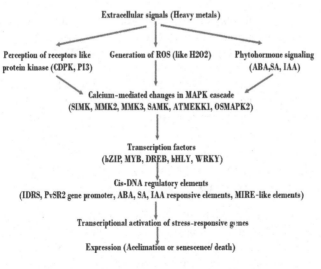

FIGURE 5.2 Signaling pathways and the molecular processes of phytoremediation for soil exposed to heavy metals.

3. **MAPK Pathways:** In all eukaryotes, they result in the transmission of signals from outside the cell to the intracellular target of the cell (Li et al., 2006). In Alfalfa, Cd and Cu trigger four distinct MAPKS: SIMK, MMK2, MMK3, and SAMK. One in Arabidopsis, i.e., ATMEKK1, and the other in rice, i.e., OSMAPK2. The CDPK (calcium-dependent protein kinases) and PI3 kinase (phosphatidyl inositol 3 kinases) are necessary for the Cd and Cu-triggered MAPK activation. Hence, Ca and Cu result in both ROS production and Ca accumulation. MAPKs link cytoplasm to the nucleus and trigger other protein kinases, TFs, and regulatory proteins (Shao et al., 2010).

4. **Phytohormone Signaling:** Abscisic acid(ABA), Salicylic acid (SA), and Indole-3-acetic acid (IAA)are involved in the response mechanism to heavy metals. Auxin-triggered mRNA was observed in *Brassica juncea* seedlings induced with Cd. In *Arabidopsis thaliana,* Cd treatment showed that the presence of nitrilase protein triggered the biosynthesis of auxin. Hg treatment results in SA methyltransferase (SAMT) gene activation that is involved in SA. CD produces ABA and ET, or other stress hormones and reactions. They may serve roles like induction of responsive genes and effectors in certain reactions related to heavy metals.

Heavy metals induce transcription factors and responsive elements. Heavy metals can activate transcription factors, which regulate the transcriptional process. The induction of Cd results in the activation of leucine zipper and zinc finger TFs (*A. thaliana* and *B. juncea*). The DREBA (dehydration responding element binding proteins) gene was upregulated by Cd, which activates the rd29 gene that plays a principal role under high salt, cold, and dehydration. Zn treatment also results in the induction of TF BHLH and reduction of WRKY and Zn finger GATA type. It is not clear yet whether the activation of TFs is specific to metal or common. In plants, the ROS-mediated process of Tf activation is a common link in the various stress responses. ROS seems to play an important role. Heavy metal-induced TFs bind to responsive elements present in the promotor of responsive genes of heavy metal (Shao et al., 2010). There are now just two kinds of cis-DNA elements known. One kind is iron-dependent regulatory sequences (IDRS), which control transcription regulated by iron and are involved in the acquisition of iron. Second is the promotor region of the PVSR2 gene provided by *Proteus vulgaris*. The PVSR2 gene encodes a heavy metal related to stress protein, and its expression is significantly increased through Hg, Cd, As, and Cu.

5.3   CHELATES AND PLANT GROWTH REGULATORS ASSISTED PHYTOEXTRACTION

One of the most effective technologies to reduce the level of heavy metals from the soil using plants is known as phytoextraction. It includes the absorption and accumulation of heavy metals into harvestable portions. Natural species of plant known as hyperaccumulators show phytoextraction, but the process is very slow due to their sluggish growth and less biomass. The process of phytoextraction can be accelerated in hyperaccumulator as well as in non-hyperaccumulator species by the use of plant growth regulators and some chelating agents. This process is known as chelate-assisted phytoextraction and PGR-assisted phytoextraction, respectively.

5.3.1 ROLE OF CHELATING AGENTS

Chelating agents are divided into two major groups: synthetic and organic/natural chelating agents.

1. **Synthetic:** Some successfully used synthetic chelating agents are EDTA (ethylene diamine tetra acetic acid), EGTA (ethylene glycol tetra acetic acid), and DTPA (diethylene triamine penta acetic acid). Application of a chelating agent results in the chelate-heavy metal complex, which increases the solubility of metals and its absorption onto plant roots. But, the major risk associated with the usage of synthetic chelating agents is related to the high mobility in the soil. High mobility results in the leaching and transport of metal from contaminated sites to uncontaminated sites. Another problem of utilizing synthetic chelating agents is that they might surpass heavy metal absorption by plants, increasing plant toxicity. Hence, special attenuation is needed toward the use of alternative strategy (Pereira et al., 2010).

2. **Organic:** Low molecular weight organic acids (LMOWA), or natural organic chelating agents, are another approach that utilizes the same idea. LMOWA form low to moderate stability with metals. A few examples are citric acid, oxalic acid, malic acid, and acetic acid. Citrate and acetate found Cd-ligands in *Solanum nigrum* (Sun et al., 2006) and citrate found Ni-ligand in *Thlaspi goesingense*. LMOWA have an advantage as they have a high rate of biodegradation than synthetic chelating agents. Also, they are not found to be toxic for plant growth. Also, the organic chelators can be applied to a wide range of heavy metals. Organic chelating agents should be applied close to the harvestable stage because organic acids are carbon sources for soil microorganisms, thus the action of microorganisms is also found to be effective.

5.3.2 ROLE OF PLANT GROWTH REGULATORS

The usage of PGR is a method for increasing the effectiveness of phytoextraction, reducing the negative impact on growing plants, and boosting the biomass of the shoots (Ouzounidov and Ilias, 2005). The two most important PGRs are auxin and cytokinin involved in assisted phytoextraction. The uptake of minerals/contaminants from roots to shoots is principally determined by the rate of transpiration. Transpiration is the driving force for the uptake of minerals/contaminants. Stomatal function, leaf area, hormone synthesis, and other biotic factors influence transpiration, as can abiotic ones such as soil water availability, climate, and water depth.

The rate of transpiration is dependent upon the opening and closing of stomata, i.e., wide open stomata account for increased transcription. Hence, the chemical regulation of stomata opening can be used to improve phytoextraction. Exogenous CKs (cytokinins) such as zeatin and kinetin can improve transpiration rate in excised leaves of *Helianthus*, *Avena*, *Brassica*, *Hordeum*, *Triticum*, and *Vigna* by significantly increasing stomatal aperture. The use of IAA can also result in improved phytoextraction as auxin is directly involved in biomass production and the overall growth of plants. Fuentus et al. (2000) treated *Zea mays* with IBA or NAA, which resulted in a 41.2% and 127.4% increase in Pb uptake. But, a huge PGR concentration can decrease the overall biomass. Some examples of plant growth regulators in phytoremediation are as follows:

1. **Auxin:** Auxin is involved in various vital metabolic processes in plants. Hence, auxin has a positive effect on the phytoextraction efficiency. Its main function is cell elongation, maturation, growth, organ differentiation, and apical dominance. Commonly detected plant auxins are IAA (indole-3-acetic acid) and IBA (indole-3-butyric acid).
2. **Cytokinins:** Cytokinin is majorly involved in cell division and is also involved in the accumulation of chlorophyll, bud formation, etc. It is also associated with the process of senescence, plant resistance salinity, and many others. Cytokinin promotes the opening of stomatal pores, which improves phytoextraction effectiveness. The higher the rate of transpiration, the more pollutants are transported from soil to plants.
3. **Gibberellins:** Gibberellins are involved in the germination of seeds, sex determination, leaf expansion, and stem elongation. External gibberellins preserve the photosynthetic system toward heavy metal toxicity (Ouzounidou and Ilias, 2005).

5.4 CONCLUSIONS

When the content of heavy metals is low, it is important for plant growth and survival, but when it is too much, it becomes very toxic and causes toxic symptoms and inhibits the growth of plants. The impact of heavy metals on plant growth and development of plants seems to be mainly caused by chromosomal abnormalities, cell abnormalities, and also due to metal-induced inhibition of the photosynthetic process. Some plants have the capacity to purify these heavy metals and convert them into less toxic forms called accumulators. Phytoaccumulators involve the usage of various molecular and biochemical methods to improve the healing process, such as PGR (plant growth hormone) and chelators that aid heavy metal absorption. Coordinated biochemical and molecular mechanisms improve the phytoremediation process, which involves the activation and expression of various stresses-responsive genes and their cascade. This results in the plant's survival and growth under stressful situations of heavy metal contamination. Thus, the process is of utmost importance for sustainable environmental conservation.

REFERENCES

Atal N., Saradhi P.P., Mohanty P., (2003). Inhibition of chloroplast photochemical reaction by treatment of wheat seedlings with low concentrations of cadmium: analysis of electron transport activities and changes in fluorescence yield. *Plant Cell Physiology*, 32: 943–951.

Baker A.J.M., Reeves R.D., Smith J.A.C., (1991). In situ decontamination of heavy metal polluted soils using crops of metal accumulating plants – A feasibility study. In: *In situ bioremediation* (Eds Hinchee R.E., Olfenbuttel R.F.), Stoneham, MA: Butterworth-Heinemann, pp. 539–544.

Baker, A.J.M., Walker P.L., (1989). Ecophysiology of metal uptake by tolerant plants. In: *Heavy metal tolerance in plants: Evolutionary aspects* (Ed Shaw A.J.), Boca Raton, FL: CRC Publication, pp. 155–177.

Baker, A.J.M., McGrath, S.P., Reeves, R.D., Smith, J.A.C., (2000) Metal hyperaccumulator plants: A review of the ecology and physiology of a biological resource for phytoremediation of metal-polluted soils. In: *Phytoremediation of Contaminated Soil and Water* (Eds Terry N., Banuelos G.), London: Lewis Publishers, pp. 85–107.

Barceló, J., Poschenrieder, C., (1990). Plant water relations as affected by heavy metal stress: a review. *Journal of Plant Nutrition*, 13: 1–37.

Barcelo J., Poschenrieder C., Gunse B., (1986). Water relations of chromium VI treated bush bean plants (*Phaseolus vulgaris* L cv Contender) under both normal and water stress conditions. *Journal of Experimental Botany*, 37: 178–187.

Beyersmann D., Hartwig A., (2008). Carcinogenic metal compounds: recent insight into molecular and cellular mechanisms. *Archives of Toxicology*, 82: 493–512.

Clarkson T.W., (1972). The pharmacology of mercury compounds. *Annual Review of Pharmacology*, 12: 375–406.

Dembitsky V., (2003). Natural occurrence of arseno compounds in plants, lichens, fungi, algal species, and microorganisms. *Plant Sciences* 165: 1177–1192.

Deng D.M., Shu W.S., Zhang J., Zou H.L., Ye Z.H., Wong M.H., Lin Z., (2007). Zinc and cadmium accumulation and tolerance in populations of Sedum alferdii. *Environmental Pollution*, 147: 381–386.

El-Sheekh M.M., (1992). Inhibition of photosystem II in the green alga *Scenedesmus obliquus* by nickel. *Journal of Biochemistry and Physiology*, 188: 363–372.

Fernandes J.C., Henriques F.S., (1991). Biochemical, physiological and structural effects of excess copper in plants. *Botanical Review*, 57: 264–273.

Fodor E., Szabo-Nagy A., Erdei L. (1995). The effects of cadmium on the fluidity and H+-ATPases activity of plasma membrane from sunflower and wheat roots. *Journal of Physiology*, 147: 87–92.

Gleba D., Bortrjuk N.V., Bortsjuk L.G., Gonden P., (1999). Use of plant roots for phytoremediation and molecular farming. *Proceedings of the National Academy of Sciences of the United States of America*, 96: 5973–5977.

Jaffre T., Pillon Y., Thomine S., Merlot S., (2013). The metal hyperaccumulators from New Caledonia can broaden our understanding of nickel accumulation in plants. *Frontiers of Plant Sciences*, 4: 1–7.

Karimi N., Ghaderian S.M., Raab A., Feldmann J., Meharg A.A., (2009). An asenic-accumulating, hypertolerant *Brassica isati, Cappadoicica. Journal of New Phytology*, 184: 41–47.

Krupa Z., Baszynsky T., (1995). Some aspects of heavy metals toxicity towards photosynthetic apparatus: direct and indirect effects on light and dark reactions. *Journal of Acta Physiologiae Plantarum*, 17: 177–190.

Li Y., Dankher O.P., Carretra L., Jtang L.X., (2006). The shoot-specific expression of gamma-glutamylcystienesynthtase directs the long distance transport oh thiol-peptides to roots conferring tolerance to mercury and arsenic. *Journal of Plant Physiology*, 141: 288–298.

Ma L.Q., Komar, K.M., Tu C., Zhang W., Cai Y., Kennelley E.D., (2001). A fern that hyperaccumulate arsenic. *Nature*, 409: 579–579.

Nedjimi B., Daoud Y., (2009). Cadmium accumulation in *Altriplex halimus* subsps chweinfurthii and its influence on growth, proline, root hydraulic conductivity and nutrient uptake. *Flora Morphology, Distribution, Functional Ecology of Plants*, 204: 316–324.

Ochiai E.L., (1987). *General principal of biochemistry of the elements*. New York: Plenum Press.

Ouariti O., Bousamma N., Zarrouk M., Cherif A., Ghorbali M.H., (1997). Cadmium and copper induced changes in tomato membranes lipids. *Journal of Phytochemistry*, 45: 1343–1350.

Ouzounidou G., Ilias I., (2005). Hormone induced protection of sunflower photosynthetic apparatus against copper toxicity. *Journal of Biology of Plant*, 49(2): 233–228.

Pereira B.F.F., De Abreu C.A., Herpin, U., De Abreu M.F., Berton R.S., (2010). Phytoremediation of lead by jack beans on a rhodicHapludox amended with EDTA. *Scientia Agricola*, 67: 308–318.

Pilon-Smits, E.A.H., Quinn C.F., Tapken W., Malagoni M., Schiavon M., (2009). Physiological functions of beneficial elements. *Current Opinnions in Plant Biology*, 12: 267–274.

Raskin I., Kumar P.BAN., Dushenkow S., Salt D.E., (1994). Bioconcentration of heavy metal by plants. *Current Opinion in Biotechnology*, 5(3): 285–290.

Reeves R.D., Baker A.J.M., (2000). Metal accumulating plants. In: *Phytoremediation of toxic metals: Using plants to clean up the environment* (Eds Raskin I., Ensley B.). New York: Wiley, pp. 193–229.

Shao et al., (2008). Physiological and molecular responses of higher plant to abiotic stresses. In: *Abiotic stress and plant responses* (Eds Khan N.A., Singh S.). New Delhi: I.K. International Publishing House Pvt. Ltd., pp. 1–22.

Shao et al., (2010). Understanding molecular mechanisms for improving phytoremediation of heavy metal-contaminated soils. *Critical Views in Biotechnology*, 30(1): 23–30.

Siedlecka A., Baszynsky T., (1993). Inhibition of electron flow around photosystem I in chloroplasts of cd-treated maize plants is due Cd-induced iron deficiency. *Journal of Physiology of Plant*, 87: 199–202.

Spinoza-Quinones F.R., Zacarkim C.E., Palacio S.M., Obregon C.L., Zennatti D.C., (2005). Removal of heavy metal from polluted river water using aquatic macrophytes *Salvinia* sp. *Braz. Journal of Physiology*, 35: 744–746.

Stochs S.J., Bagchi D., (1995). Oxidative mechanism in the toxicity of metals ions. *Free Radical Biology and Medicine*, 18: 321–336.

Sun R.L., Zhou Q.X., Jin C.X., (2006). Cadmium accumulation in relation to organic acids in leaves of Solanumnigrum L. as a newly found cadmium hyperaccumulator. *Plant and Soil*, 285: 125–134.

Suresh B., Ravishankar G.A., (2004). Phytoremediation a novel and promising approach for environmental clean up. *Critical Reviews in Biotechnology*, 24: 97–124.

Tamaoki M., Freeman J.L., Pilon-Smits E.A.H., (2008). Coopoerative ethylene and jasmonic acid signaling regulates selenite resistance in Arabidopsis. *Journal of Plant Physiology*, 146:1219–1230.

UNEP. (undated). *Phytoremediation: an environmentally sound technology for pollution prevention, control and remediation. An introductory guide to decision-makers. Newsletter and Technical Publications Freshwater Management Series No. 2, United Nations Environment Programme Division of Technology, Industry, and Economics.*

Vajpayee P., Rai U.N., Ali M.B., Tripathi R.D., Yadav V., Sinha S.N. (2001). Chromium inuced physiological changes in Vallisneriaspiralis L. and its role in pytoremediation of tannery effluent. *Bulletin of Environmental Contamination and Toxicology*, 67: 246–256.

Yang, X., Feng, Y., He, Z., Stoffella, P., (2005). Molecular mechanisms of heavy metal hyperaccumulation and phytoremediation. *Journal of Trace Elements in Medicine and Biology: Organ of the Society for Minerals and Trace Elements (GMS)*, 18: 339–353. https://doi.org/10.1016/j.jtemb.2005.02.007.

Zhang X., Xia H., Li Z., Zhang P., Gao B., (2010). Potential of four forage grasses in remediation of cd and Zn contaminated soils. *Bioresource Technology*, 101: 2063–2066. https://doi.org/10.1016/j.biortech.2009.11.065.

Zhen-Guo S., Xian-Dong L., Chun-Chun W., Huai-Man Ch., Hong Ch., (2002). Lead Phytoextraction from contaminated soil with high-biomass plant species. *Journal of Environmental Quality*, 31: 1893–1900.

6 Computational Modelling in Bioremediation
Innovations and Future Directions

Adarsh Kumar Shukla, Ashwani Kumar,
Anamika, Pradeep Kumar, and Ankush Yadav

6.1 INTRODUCTION

The bacteria, plants, or their enzymes are widely used to remove contaminants from environmental surfaces; this phenomenon is termed as bioremediation. This innovative technology can be applied alongside conventional approaches based on physical and chemical methods to address the intricate gathering of pollutants (Ahmed et al., 2021). It appears to be a sustainable method of managing environmental contamination; thus, advances and more research studies are needed in this area. The elimination of complex environmental pollutants for the safety of humans and the environment requires improvement, as conventional approaches to bacterial-based bioremediation are time-consuming, expensive, and less impressive (Patel et al., 2022). This type of bioremediation needs some advances to its intermediate steps that speed up the overall process. Pollutants from paper mills are chemically phenolic and chlorinated, and because of this, they pose a serious risk to both the environment and human health (Lindholm-Lehto et al., 2015). Several studies have suggested that bioremediation could be greatly accelerated by using *in silico* methods. A recent development is the field of "*in silico* bioremediation", which looks for innovative approaches to use microbial enzymatic systems to improve and enhance the traditional bioremediation process of pollutants from paper mills (Singh et al., 2020). Computational approaches in bioremediation techniques are being used to remove harmful substances for a safe and sustainable environment in order to achieve impactful results. Numerous software programs, applications, and database resources are accessible for conducting investigations to support the bioremediation process through the integration and analysis of omics datasets. The integration of "genomics, transcriptomics, proteomics, metabolomics, and other molecular data" is a key player in bioremediation (Verma et al., 2022). Some techniques include mathematical algorithms, molecular modelling, molecular docking, and molecular dynamics simulation (Singh et al., 2020). The core of *in silico* bioremediation is molecular modelling and molecular docking simulations. These allow for the rapid catalysis simulation of contaminants and enzymes on a computer system, thereby maximizing the catalytic potential of enzymes (Singh et al., 2020).

DOI: 10.1201/9781003407317-6

6.2　MOLECULAR MODELLING

Computational molecular modelling is the process of studying the composition, characteristics, and role of molecules through the use of mathematical methods and computer simulations. This is an interdisciplinary field that incorporates elements from computer science, physics, biology, and chemistry (Badar et al., 2022). It is extensively utilized in material science, drug discovery, bioremediation, and the molecular understanding of biological processes (Singh et al., 2020). Several techniques used for molecular modelling are illustrated in Figure 6.1.

6.2.1　THREE-DIMENSIONAL STRUCTURE PREDICTION OF CHEMICAL COMPOUNDS

It is crucial in the bioremediation process to predict the three-dimensional (3D) structure of chemical compounds (pollutants). These substances constitute the fundamental configuration of the enzyme-ligand docking mechanism (Yadav et al., 2024), which is advantageous in comprehending the degradation pathways involved in bioremediation. Further investigations into virtual screening, bioactivity prediction, and rational drug design are necessary in order to obtain a dependable 3D structure for molecules (Axen et al., 2017). Utilization of robust tools, access to computational resources, and knowledge of computational chemistry and molecular modelling may be necessary for the bioremediation procedures (Tiwari and Singh 2022). Some very common techniques used for the three-dimensional structure prediction of compounds are given in Figure 6.2.

6.2.2　STEPS OF THREE-DIMENSIONAL STRUCTURE PREDICTION OF COMPOUNDS

Here are the steps involved in predicting the three-dimensional (3D) structure of compounds:

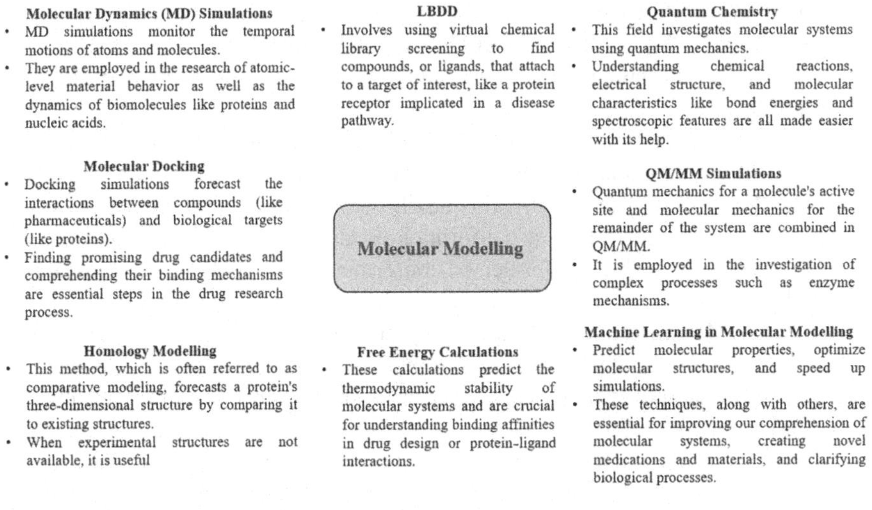

Molecular Dynamics (MD) Simulations
- MD simulations monitor the temporal motions of atoms and molecules.
- They are employed in the research of atomic-level material behavior as well as the dynamics of biomolecules like proteins and nucleic acids.

Molecular Docking
- Docking simulations forecast the interactions between compounds (like pharmaceuticals) and biological targets (like proteins).
- Finding promising drug candidates and comprehending their binding mechanisms are essential steps in the drug research process.

Homology Modelling
- This method, which is often referred to as comparative modeling, forecasts a protein's three-dimensional structure by comparing it to existing structures.
- When experimental structures are not available, it is useful

LBDD
- Involves using virtual chemical library screening to find compounds, or ligands, that attach to a target of interest, like a protein receptor implicated in a disease pathway.

Molecular Modelling

Free Energy Calculations
- These calculations predict the thermodynamic stability of molecular systems and are crucial for understanding binding affinities in drug design or protein-ligand interactions.

Quantum Chemistry
- This field investigates molecular systems using quantum mechanics.
- Understanding chemical reactions, electrical structure, and molecular characteristics like bond energies and spectroscopic features are all made easier with its help.

QM/MM Simulations
- Quantum mechanics for a molecule's active site and molecular mechanics for the remainder of the system are combined in QM/MM.
- It is employed in the investigation of complex processes such as enzyme mechanisms.

Machine Learning in Molecular Modelling
- Predict molecular properties, optimize molecular structures, and speed up simulations.
- These techniques, along with others, are essential for improving our comprehension of molecular systems, creating novel medications and materials, and clarifying biological processes.

FIGURE 6.1　Techniques involved in computational molecular modelling.

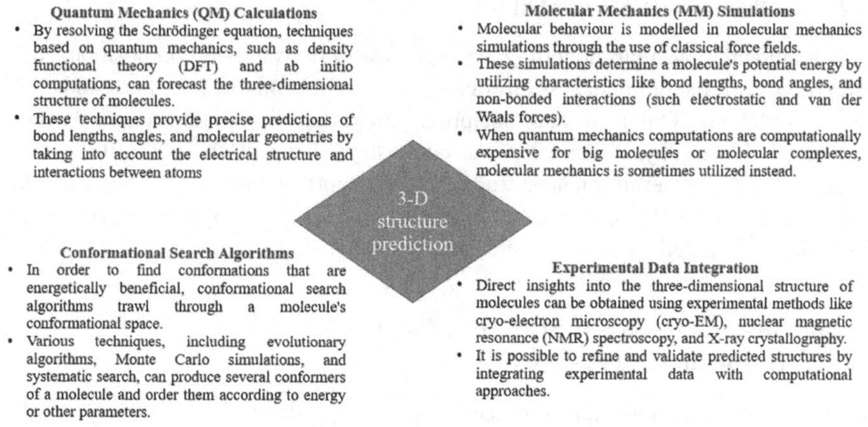

Quantum Mechanics (QM) Calculations
- By resolving the Schrödinger equation, techniques based on quantum mechanics, such as density functional theory (DFT) and ab initio computations, can forecast the three-dimensional structure of molecules.
- These techniques provide precise predictions of bond lengths, angles, and molecular geometries by taking into account the electrical structure and interactions between atoms

Molecular Mechanics (MM) Simulations
- Molecular behaviour is modelled in molecular mechanics simulations through the use of classical force fields.
- These simulations determine a molecule's potential energy by utilizing characteristics like bond lengths, bond angles, and non-bonded interactions (such electrostatic and van der Waals forces).
- When quantum mechanics computations are computationally expensive for big molecules or molecular complexes, molecular mechanics is sometimes utilized instead.

3-D structure prediction

Conformational Search Algorithms
- In order to find conformations that are energetically beneficial, conformational search algorithms trawl through a molecule's conformational space.
- Various techniques, including evolutionary algorithms, Monte Carlo simulations, and systematic search, can produce several conformers of a molecule and order them according to energy or other parameters.

Experimental Data Integration
- Direct insights into the three-dimensional structure of molecules can be obtained using experimental methods like cryo-electron microscopy (cryo-EM), nuclear magnetic resonance (NMR) spectroscopy, and X-ray crystallography.
- It is possible to refine and validate predicted structures by integrating experimental data with computational approaches.

FIGURE 6.2 Techniques involved in three-dimensional structure prediction of compounds.

i. Data Preparation: Start with the chemical structure of the compound, either drawn using molecular drawing software or obtained from databases (Deepika et al., 2023). Convert the structure into a suitable format for computational modelling, such as "Simplified Molecular Input Line Entry System (SMILES)" notation or "structure data file (SDF)" format (Thareja et al., 2021).

ii. Geometry Optimization: To maximize the compound's shape, use computational techniques such as "molecular mechanics (MM)" simulations or "quantum mechanics (QM)" calculations (van der Kamp and Mulholland, 2013). MM approaches, such as force fields like "Assisted Model Building with Energy Refinement (AMBER), Chemistry at HARvard Macromolecular Mechanics (CHARMM), or Optimized Potentials for Liquid Simulations (OPLS)", are faster but less accurate for some compounds, while QM methods, such as "density functional theory (DFT)", provide correct electronic structure information but are computationally demanding (Senn and Thiel, 2009).

iii. Conformational Sampling: Use Monte Carlo approaches, systematic search algorithms, or molecular dynamics (MD) simulations for the conformational sampling if the chemical is flexible or has numerous conformations. Create a range of different conformations and then choose the most representative or stable conformer or conformers for additional examination (Hawkins, 2017).

iv. Energy Minimization: Perform energy minimization to refine the selected conformations and achieve a more stable structure. Adjust parameters such as force field settings, convergence criteria, and solvent effects if applicable.

v. Validation and Quality Assessment: Validate the predicted structures using experimental data (if available) or computational validation methods. Compare predicted structural properties (e.g., bond lengths, angles, and torsion angles) with experimental values or known databases to assess the quality of the predictions (Rai et al., 2022).

vi. Visualization and Analysis: Visualize the 3D structure using molecular visualization software (e.g., PyMOL, VMD, and UCSF Chimera) to analyse molecular features and interactions (Yadav et al., 2021). Analyse the key structural elements, such as functional groups, bond types, and spatial arrangements, to understand the compound's properties and behaviour.

6.2.3 DETERMINATION OF 3D STRUCTURE OF ENZYMES OR PROTEINS

Three-dimensional (3D) structure prediction is a crucial aspect of computational molecular modelling, especially in the context of proteins and other biomolecules. The best strategy will rely on a number of variables, including the target molecule's (protein or enzyme) size and complexity, computer capabilities, and experimental data availability (Amaro and Mulholland, 2018). The topic of 3D structure prediction is being advanced by ongoing research in computational biology and bioinformatics, which will improve its accuracy and use for a variety of molecular systems. Some key methods used for predicting the three-dimensional structures of proteins are depicted in Figure 6.3.

6.2.3.1 Steps of 3D Structure Prediction of Proteins/Enzymes

Here are the general steps involved in predicting the three-dimensional (3D) structure of compounds computationally: Generally, the three most popular approaches, homology modelling, ab initio modelling, and fold recognition, have several tools. For homology modelling, MODELLER, SWISS-MODEL, and Phyre2 are very popular tools (Figure 6.4). For ab initio modelling, Rosetta and AlphaFold are highly used tools. I-TASSER is an online tool used for fold recognition and threading.

Homology Modelling (Comparative Modelling)
- Using the known structure of a homologous protein as a basis, this approach predicts the three-dimensional structure of a target protein.
- It is predicated on the idea that proteins with comparable structural similarities typically have comparable structures and activities.
- When the target protein's experimental structure is unknown but similar structures are known, homology modelling is especially helpful.

Ab Initio Protein Structure Prediction
- Without using pre-existing templates, ab initio approaches forecast protein architectures using empirical energy functions and physical principles.
- These techniques search for the lowest energy state, which is associated with the most stable structure, by using algorithms to explore the protein's conformational space.
- However, because of the intricacy of the conformational search, ab initio prediction is computationally demanding and frequently restricted to smaller proteins.

Fragment-based Methods
- Using fragment-based techniques, the protein structure is divided into smaller pieces, or building blocks, which are then put together to create the final three-dimensional structure.
- These techniques integrate computational algorithms with actual data—such as data from cryo-electron microscopy (cryo-EM) or NMR spectroscopy—to improve and validate the projected structure.

Machine Learning-based Structure Prediction
- The application of machine learning technologies, including deep learning models, to the prediction of protein structures is growing.
- Large databases of known protein structures may teach these models intricate patterns and correlations, which helps them predict novel protein sequences with accuracy.

Hybrid Methods
- To increase the precision and dependability of structure predictions, hybrid approaches combine many computational techniques, such as ab initio methods with experimental data or homology modelling with molecular dynamics simulations.
- The quality of projected structures can be improved, for instance, by incorporating experimental limitations such as distance limits or secondary structure predictions

3-D structure prediction of biomolecules

FIGURE 6.3 Commonly used approaches to the three-dimensional structure prediction of proteins/enzymes.

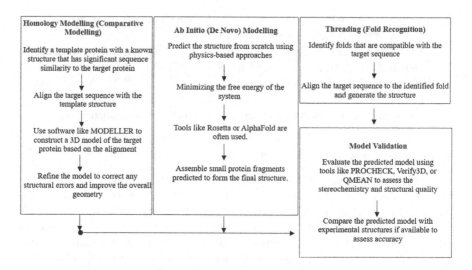

FIGURE 6.4 Pipeline of the three-dimensional structure prediction of enzymes or proteins.

The predicted model should be validated by the use of several online tools such as PROCHECK, Ramachandran plot, Verify3D, and Qualitative Model Energy ANalysis (QMEAN) (Bogra et al., 2024). After obtaining the successful three-dimensional structure of proteins and enzymes, refinement of the predicted structure is necessary. Some force fields very common for structure refinement areGROningen MAchine for Chemical Simulations (GROMACS), AMBER, and CHARMM.

6.3 COMPUTATIONAL MAPPING OF ENZYME-POLLUTANT INTERACTION

A computer technique called "molecular docking" is used to predict a ligand's preferred orientation, binding site confirmation, and binding affinity to a macromolecular target, particularly a protein (Vishal et al., 2022). The intermolecular interactions between the hazardous compounds against the biomolecules such as enzymes, proteins, nucleic acids, and lipids, play a crucial role in understanding their potential toxicological effects and mechanisms of action within biological systems. Molecular docking aims to achieve an ideal shape for the suitable protein-compound complex in order to lower the free energy of the overall system (Tripathi & Misra, 2017). For many years, *in silico* screening techniques have offered a practical and affordable option throughout the initial stages of the bioremediation process. In particular, molecular docking provides useful information for vast chemical libraries quickly and has several success stories to support the method. It approximates the enzyme-chemical/enzyme-enzyme interaction through the binding score and mapping of bondings between them (Varela-Rial et al., 2022). This approach is useful to find out the best complimentary action of chemical pollutants against the 3D structure of the target enzymes through the bio-affinity profile (Tripathi & Misra, 2017) (Figure 6.5). Investigation of chemical pollutant-enzyme interactions mapping is a

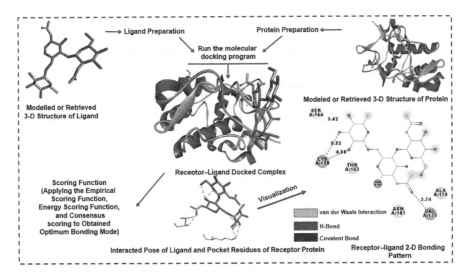

FIGURE 6.5   A graphical illustration of the molecular docking working protocol.

topic of great interest since it offers a better understanding of molecular recognition, interaction mechanism, and subsequent impact.

Numerous parts of molecular docking, such as the preliminary phase of docking stages, various inter- and intra-molecular interactions, and software packages with respective algorithms and their applications are given in Table 6.1. For the pharmaceutical business, molecular docking research is essential for identifying disease targets and creating potent medications. There are three main subcategories of the molecular docking methodology.

6.3.1   DOCKING WITH AN INDUCED-FIT APPROACH

Induced docking is the approach where both the ligand and receptor are adaptable to maximize the forces holding the ligand and receptor together. In this method, the ligand freely binds at the "active site receptor chain". It carries out the idea of protein and ligand complementarity (Blanes-Mira et al., 2022). The protein is inflexible, and only one of the molecules (chemical compound) is movable. As a result, in addition to the "six translational and two rotational" degrees of freedom (DOF), the ligand's structural DOF are changed accordingly. These techniques assume that a protein's stable conformation may be the same as the one that can recognize the ligands to be docked (Shukla and Kumar, 2023).

6.3.2   DOCKING WITH THE LOCK-AND-KEY APPROACH

This approach is based on the "lock-and-key theory". In this theory, both the chemical molecule (ligand) and the protein (receptor chain) exhibit rigidity and tight binding. It explains the fundamental idea behind three-dimensional complementarity (Jain et al., 2024). In the case of too many conformational DOF such as "protein-protein docking", this binding model (lock-key) approximation is widely used.

TABLE 6.1

List of Widely Popular Molecular Docking Tools with their Details

Sr. No.	Program and Algorithm	Description	Year of Release	Organization	Licence
1	"AutoDock and AutoDock Vina"	• Small ligands and large binding sites. • Accuracy in highly flexible ligands.	1990	"The Scripps Research Institute"	Open source (GNU GPL)
2	DOCK	• Small binding sites, opened cavities, and small hydrophobic ligands.	1988	"University of California, San Francisco"	Freeware for academic use
3	FlexAID	• Target side-chain flexibility and soft scoring function. • Based on surface complementary.	2015	"University of Sherbrooke"	Open source (Apache Licence)
4	Glide	• Flexible ligands and small hydrophobic ligands.	2004	"Schrödinger, Inc."	Commercial
5	GOLD	• Small binding sites. • Small hydrophobic ligands and buried binding pockets.	1996	"Cambridge Crystallographic Data Center"	Commercial free evaluation (2 months)
6	Surflex	• Large and opened cavities. • Small binding sites and very flexible ligands.	2003	"University of California"	Commercial package
7	Molecular Operating Environment (MOE)	• Docking application within MOE. • Choice of placement methods (including alpha sphere methods) and scoring functions (including London dG).	2008	"Chemical Computing Group"	Commercial
8	rDock	• HTVS of small molecules against proteins and nucleic acids. • Binding mode prediction.	1998	"Vernalis R&D (commercial)", "University of York (academic)", and "University of Barcelona (open source)"	Open source (GNU LGPL) (formerly commercial and academic)

(*Continued*)

TABLE 6.1 (Continued)
List of Widely Popular Molecular Docking Tools with their Details

Sr. No.	Program and Algorithm	Description	Year of Release	Organization	Licence
9	SEED	• Automated docking of fragments with evaluation of free energy of binding, including electrostatic solvation effects in the continuum dielectric approximation (generalized born).	1999	"University of Zurich"	Open source (GNU GPL)
10	LeDock	• Program for fast and accurate flexible docking of small molecules into a protein.	2016	Lephar	Freeware for academic use
11	Glide	• Systematic search of the conformations. • Orientations and positions of the docked ligand.	2004	"Schrödinger, Inc."	Proprietary software

6.3.3 ENSEMBLE DOCKING

This method states the adaptability and complexity of protein "conformational states". For docking with a ligand, several protein structures are used as an ensemble. Manual molecular docking and automatic molecular docking are possible for the virtual screening of lead molecules (Shukla and Kumar, 2023). In manual docking, ligands are coupled with their corresponding groups in the binding site since the binding groups on the ligand and binding site are known (Shukla and Kumar, 2024). Each potential interaction has a predetermined bonding distance. To achieve the optimal match as specified by the operator, the program positions the chemical compound (ligand molecule) around the binding residues of the receptor chain (Pandit et al., 2023).

6.3.4 MECHANISM OF MOLECULAR DOCKING

Numerous robust tools are available for mapping the protein-chemical interactions. The AutoDock Vina is a freely available command-based approach widely used for determining the binding pattern between the enzyme or protein and chemical compounds (Shukla and Kumar, 2023). This program performs a series of steps (Figure 6.6) for the molecular docking and mapping of bondings between the protein and chemical compounds.

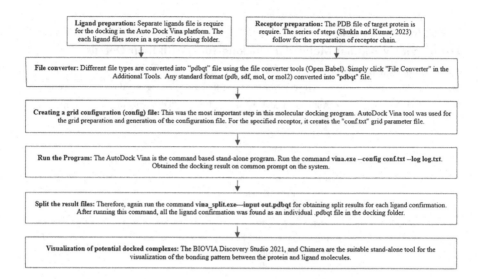

FIGURE 6.6 Pipeline of the molecular docking approach in the AutoDock Vina platform.

6.3.4.1 Docking Score

The molecular docking quickly and precisely evaluates protein-ligand complexes or approximates the interaction's energy. To rank these complexes and distinguish between predictions of valid binding modes and predictions of invalid binding modes, an effective scoring function is required (Yadav et al., 2024). A chemical entity and protein docking method may produce a huge number of target-ligand complexes, which forecast the binding affinities between them. There are four categories of scoring functions of docking such as "consensus score, empirical score, knowledge-based score, and force field or molecular mechanics-based score".

6.3.5 Advantages of Molecular Mapping

Molecular docking is a well-known *in silico* approach that is often used in drug development procedures. Finding novel therapeutic compounds and foreseeing "ligand-target interactions" at the molecular level is made possible by docking since it can be done without prior knowledge of the chemical composition of other target modulators. Despite being first created to aid in studying the mechanics of molecular mapping between small chemical compounds and large proteins, docking's functions and applications in drug development have undergone significant modification in recent years (Pinzi and Rastelli, 2019). The capacity of molecular docking to effectively and efficiently identify novel enzymes or proteins against the chemical pollutants from a huge chemical library is its most significant advantage. Typically, novel biomolecules for a particular target that are unrelated to recognized ligands may have new biological effects.

Zhu et al. (2022) carefully examined the docking performance against new targets and the structural uniqueness of the docking hits to comprehend this benefit of docking screening. The docking results should be verified using "structural, biochemical, cellular function, and in vivo studies". Zhu et al. (2022) reviewed 252 docking results and found that 17.1% of the studies determined the three-dimensional structures, 65.8% of the studies conducted cellular experiments, and only 9.5% of the studies performed animal tests.

6.3.6 RECENT SUCCESS STORIES OF MOLECULAR DOCKING-BASED BIOREMEDIATION

In the realm of bioremediation, molecular docking is an effective method that makes it easier to identify, optimize, and apply biological agents to remove environmental pollutants. By simulating how pollutants interact with naturally existing enzymes and bacteria, molecular docking aids in the prediction of the fate of pollutants in the environment (Singh et al., 2021a, b). Some new stories of molecular docking-based bioremediation are given below:

i. Heavy Metal Bioremediation: Proteins that can chelate heavy metals have been identified through docking studies, which facilitate their removal from contaminated locations (Witkowska et al., 2021). The importance of introducing a hydroxamate siderophore from *Aspergillus nidulans* to *Bacillus subtilis* cells for the bioremediation of cadmium salt was empirically studied in the current study. This molecular docking-based study examined the siderophore-mediated intracellular Cd accumulation by bacterial cells, growth assessment, and biochemical tests such as total protein content, carbohydrate content, and iron content estimation (Khan et al., 2020).

ii. Organic Pollutants: Molecular docking studies have been conducted on laccases and peroxidases to improve their capacity to break down complex organic pollutants like polycyclic aromatic hydrocarbons (PAHs) (Liu et al., 2018).

iii. Pesticide Degradation: Enzymes that can efficiently break down pesticides, lowering their toxicity and persistence in the environment, have been found via docking. The study published by Bhatt et al. (2023) helped to understand how laccase catalyses the breakdown of glyphosate, isoproturon, lignin polymer, and parathion during detoxification. Using molecular docking and molecular dynamics simulation techniques, we investigated the relationships between laccase-glyphosate, laccase-lignin polymer, laccase-isoproturon, and laccase-parathion.

CONCLUSIONS

Molecular docking is the *in silico* tool that easily determines which proteins or metabolic pathways are necessary to break down particular contaminants in microorganism-based bioremediation. This can expedite the microbial community selection process for environmental applications. By simulating how pollutants interact with naturally existing enzymes and bacteria, this chapter facilitates the prediction of the fate of pollutants in the environment. This can help with risk evaluations and the creation of remediation strategies.

REFERENCES

Ahmed, S. F., Mofijur, M., Nuzhat, S., Chowdhury, A. T., Rafa, N., Uddin, M. A.,... & Show, P. L. (2021). Recent developments in physical, biological, chemical, and hybrid treatment techniques for removing emerging contaminants from wastewater. *Journal of Hazardous Materials*, 416, p. 125912.

Amaro, R. E., & Mulholland, A. J. (2018). Multiscale methods in drug design bridge chemical and biological complexity in the search for cures. *Nature Reviews Chemistry*, 2(4), p. 0148.

Axen, S. D., Huang, X. P., Cáceres, E. L., Gendelev, L., Roth, B. L., & Keiser, M. J. (2017). A simple representation of three-dimensional molecular structure. *Journal of Medicinal Chemistry*, 60(17), pp. 7393–7409.

Badar, M. S., Shamsi, S., Ahmed, J., & Alam, M. A. (2022). Molecular dynamics simulations: concept, methods, and applications. In M. A. Rahmandoust (ed.), *Transdisciplinarity* (pp. 131–151). Cham: Springer International Publishing.

Bhatt, P., Bhatt, K., Chen, W. J., Huang, Y., Xiao, Y., Wu, S.,... & Chen, S. (2023). Bioremediation potential of laccase for catalysis of glyphosate, isoproturon, lignin, and parathion: molecular docking, dynamics, and simulation. *Journal of Hazardous Materials*, 443, p. 130319.

Blanes-Mira, C., Fernández-Aguado, P., de Andrés-López, J., Fernández-Carvajal, A., Ferrer-Montiel, A., & Fernández-Ballester, G. (2022). Comprehensive survey of consensus docking for high-throughput virtual screening. *Molecules*, 28(1), p. 175.

Bogra, P., Shukla, A. K., Panwar, S., Kumar, A., Singh, J., & Singh, H. (2024). Inhibition profiling of Bestatin against the aminopeptidase B: in silico and in vitro approach. *Applied Biochemistry and Microbiology*, 60(1), pp. 124–131.

Deepika, Shukla, A. K., Kumari, A., & Kumar, A. (2023). Gut brain regulation using psychobiotics for improved neuropsychological illness. *Developmental Psychobiology*, 65(5), p. e22404.

Hawkins, P. C. (2017). Conformation generation: the state of the art. *Journal of Chemical Information and Modeling*, 57(8), pp. 1747–1756.

Jain, S., Shukla, A. K., Panwar, S., Kumar, R., & Kumar, A. (2024). In vitro antibacterial activity of antibiotics and plant essential oils against Escherichia coli MTCC443 supported through the molecular docking and pharmacokinetics study. *Biotechnology and Applied Biochemistry*, 71(4), pp. 868–880.

Khan, A., Gupta, A., Singh, P., Mishra, A. K., Ranjan, R. K., & Srivastava, A. (2020). Siderophore-assisted cadmium hyperaccumulation in Bacillus subtilis. *International Microbiology*, 23, pp. 277–286.

Lindholm-Lehto, P. C., Knuutinen, J. S., Ahkola, H. S., & Herve, S. H. (2015). Refractory organic pollutants and toxicity in pulp and paper mill wastewaters. *Environmental Science and Pollution Research*, 22, pp. 6473–6499.

Liu, Z., Liu, Y., Zeng, G., Shao, B., Chen, M., Li, Z.,... & Zhong, H. (2018). Application of molecular docking for the degradation of organic pollutants in the environmental remediation: a review. *Chemosphere*, 203, pp. 139–150.

Pandit, A., Shukla, A. K., Deepika, Vaidya, D., Kumari, A., & Kumar, A. (2023). In vitro assessment of anti-microbial activity of aloe vera (Barbadensis miller) supported through computational studies. *Russian Journal of Bioorganic Chemistry*, 49(2), pp. 342–351.

Patel, A. K., Singhania, R. R., Albarico, F. P. J. B., Pandey, A., Chen, C. W., & Dong, C. D. (2022). Organic wastes bioremediation and its changing prospects. *Science of the Total Environment*, 824, p. 153889.

Pinzi, L., & Rastelli, G. (2019). Molecular docking: shifting paradigms in drug discovery. *International Journal of Molecular Sciences*, 20(18). https://doi.org/10.3390/ijms20184331.

Rai, B. K., Sresht, V., Yang, Q., Unwalla, R., Tu, M., Mathiowetz, A. M., & Bakken, G. A. (2022). Torsionnet: a deep neural network to rapidly predict small-molecule torsional energy profiles with the accuracy of quantum mechanics. *Journal of Chemical Information and Modeling*, 62(4), pp. 785–800.

Senn, H. M., & Thiel, W. (2009). QM/MM methods for biomolecular systems. *Angewandte Chemie International Edition*, 48(7), pp. 1198–1229.

Shukla, A. K., & Kumar, A. (2023). Virtual screening of orally active lead compounds of pearl millet and their structural activity against target protein of COVID-19. *Russian Journal of Bioorganic Chemistry*, 49(Suppl 1), pp. S53–S70.

Shukla, A. K., & Kumar, A. (2024). Comparative study between antiviral drugs and natural compounds against "Cryo-EM structure of the SARS-CoV-2 Omicron spike". *Vegetos*, 37(2), pp. 596–605.

Singh, A. K., Bilal, M., Iqbal, H. M., & Raj, A. (2021a). Trends in predictive biodegradation for sustainable mitigation of environmental pollutants: recent progress and future outlook. *Science of the Total Environment*, 770, p. 144561.

Singh, A. K., Bilal, M., Iqbal, H. M., Meyer, A. S., & Raj, A. (2021b). Bioremediation of lignin derivatives and phenolics in wastewater with lignin modifying enzymes: status, opportunities and challenges. *Science of the Total Environment*, 777, p. 145988.

Singh, A. K., Chowdhary, P., & Raj, A. (2020). In silico bioremediation strategies for removal of environmental pollutants released from paper mills using bacterial ligninolytic enzymes. In P. Chowdhary, A. Raj, D. Verma, & Y. Akhter (eds.), *Microorganisms for sustainable environment and health* (pp. 249–285). Amsterdam: Elsevier.

Thareja, R., Singh, J., & Bansal, P. (2021). Computational tools in cheminformatics. In *Chemoinformatics and bioinformatics in the pharmaceutical sciences* (pp. 105–137). Academic Press.

Tiwari, A., & Singh, S. (2022). Computational approaches in drug designing. In *Bioinformatics* (pp. 207–217). Academic Press.

Tripathi, A., & Misra, K. (2017). Molecular docking: a structure-based drug designing approach. JSM Chem, 5(2), p. 1042.

van der Kamp, M. W., & Mulholland, A. J. (2013). Combined quantum mechanics/molecular mechanics (QM/MM) methods in computational enzymology. *Biochemistry*, 52(16), pp. 2708–2728.

Varela-Rial, A., Majewski, M., & De Fabritiis, G. (2022). Structure based virtual screening: fast and slow. *Wiley Interdisciplinary Reviews: Computational Molecular Science*, 12(2), pp. 1–17. https://doi.org/10.1002/wcms.1544.

Verma, S., Kour, S., & Pathak, R. K. (2022). In Silico Approaches in Bioremediation Research and Advancements. In D. C. Suyal & R. Soni (eds.), *Bioremediation of environmental pollutants: emerging trends and strategies* (pp. 221–238). Cham: Springer.

Vishal, Banyal, S., Shukla, A. K., Kumari, A., Kumar, A., Khatak, A., Luthra, A., Sunil, & Kumar, M. (2022). Effect of Modification on Quality Parameters of Jackfruit (Atrocarpus heterophyllus) Seed Starch to Valorize its Food Potential and In-Silico Investigation of the Pharmacological Compound Against Salmonellosis. Waste and Biomass Valorization. https://doi.org/10.1007/s12649-022-01945-0.

Witkowska, D., Słowik, J., & Chilicka, K. (2021). Heavy metals and human health: possible exposure pathways and the competition for protein binding sites. *Molecules*, 26(19), p. 6060.

Yadav, P., Shukla, A. K., Dhewa, T., & Kumar, A. (2021). In silico investigation of antioxidant interaction and effect of probiotic fermentation on antioxidant profiling of pearl millet-based rabadi beverage. *Journal of Applied and Natural Science*, 13(4), pp. 1531–1544.

Yadav, P., Shukla, A. K., Kumari, A., Dhewa, T., & Kumar, A. (2024). Nutritional evaluation of probiotics enriched rabadi beverage (PERB) and molecular mapping of digestive enzyme with dietary fibre for exploring the therapeutic potential. *Food and Humanity*, 2, p. 100221.

Zhu, H., Zhang, Y., Li, W., & Huang, N. (2022). A comprehensive survey of prospective structure-based virtual screening for early drug discovery in the past fifteen years. *International Journal of Molecular Sciences*, 23(24). https://doi.org/10.3390/ijms232415961.

7 Navigating Nanomaterials
A Path to Sustainable Wastewater Treatment

Love Singla

7.1 INTRODUCTION

The advent of industrialization and urbanization has significantly increased the demand for clean water while also contributing to severe water pollution. Although effective to a certain extent, traditional wastewater treatment methods often need to be revised to address the growing complexity and volume of contaminants in wastewater. This has led to exploring innovative technologies, including nanotechnology, which is a promising solution. Due to their exceptional physical and chemical properties at the nanoscale, nanomaterials have gained considerable attention in wastewater treatment. These materials exhibit a high surface area-to-volume ratio, increased reactivity, and functionalization ability for specific applications (Zeng et al., 2021). Such characteristics make nanomaterials highly efficient in removing many pollutants, including heavy metals, organic compounds, and pathogens, which are often challenging to eliminate using conventional methods (Lazarenko et al., 2022).

One of the primary mechanisms through which nanomaterials enhance wastewater treatment is adsorption. Nano-adsorbents, such as carbon nanotubes (CNTs) and metal-organic frameworks (MOFs), possess a vast surface area that effectively captures contaminants. These materials can be tailored to target specific pollutants, enhancing the selectivity and efficiency of the treatment process (Chandran et al., 2023). Additionally, nanomaterials like titanium dioxide (TiO_2) are utilized in photocatalysis, which harnesses light energy to degrade organic pollutants into less harmful substances (Canbaz et al., 2019). Nanofiltration is another crucial application of nanotechnology in wastewater treatment. Nanomaterial-based membranes exhibit superior permeability and selectivity compared to traditional membranes. These membranes can effectively remove microscopic contaminants, including viruses and bacteria, while maintaining high flow rates. Moreover, their resistance to fouling, a common issue in membrane technology, extends their operational lifespan and reduces maintenance costs (Bandehali et al., 2020).

The disinfection capabilities of nanomaterials also play a vital role in enhancing wastewater treatment. Nanoparticles, such as silver and copper, have potent antimicrobial properties that can inactivate a broad spectrum of pathogens. This antimicrobial action is particularly beneficial in preventing the spread of waterborne diseases and ensuring the safety of treated water (Kamyab et al., 2023). Furthermore, integrating nanomaterials in sensors and monitoring systems allows for real-time detection of contaminants, enabling more efficient management and treatment of wastewater (El-Sayyad et al., 2020; Oun et al., 2020).

Despite the promising potential of nanomaterials, their application in wastewater treatment is challenging. Concerns regarding nanomaterials' environmental fate and potential toxicity necessitate comprehensive risk assessments and the establishment of regulatory frameworks to ensure their safe use. Additionally, the scalability of nanomaterial production and the associated costs remain significant barriers to widespread adoption. Nanotechnology offers a transformative approach to wastewater treatment, addressing the limitations of conventional methods and paving the way for more sustainable water management practices. Continued research and development in this field are essential to fully harness the capabilities of nanomaterials and integrate them effectively into existing wastewater treatment systems.

7.2 CLASSIFICATION OF NANOMATERIALS IN WASTEWATER TREATMENT

Nanomaterials have revolutionized wastewater treatment technologies with their unique physicochemical properties. Their classification is primarily based on their functional roles and material composition. Each type offers unique advantages and pollutant removal mechanisms, contributing to the efficiency and effectiveness of wastewater treatment processes (Zamel et al., 2023). As research and development continue, the potential for new and improved nanomaterials and hybrid systems will likely expand, further advancing the field of sustainable water management. In wastewater treatment, nanomaterials are broadly categorized into three main types: nano-adsorbents, nano-catalysts, and nano-membranes, as shown in Figure 7.1.

7.2.1 NANO-ADSORBENTS

Nano-adsorbents are nanomaterials designed to remove contaminants from wastewater through adsorption processes. These materials have a high surface area-to-volume ratio, allowing them to capture and retain many pollutants (Lemessa et al., 2022). Common nano-adsorbents include:

    a. Carbon-based Nanomaterials: This category encompasses CNTs, graphene, and activated carbon. CNTs and graphene have exceptional adsorption capacities due to their extensive surface areas and porous structures, making them highly effective in removing organic pollutants, heavy metals, and dyes from wastewater (Chadha et al., 2022).

    b. Metal-organic Frameworks (MOFs): MOFs are crystalline materials of metal ions coordinated to organic ligands, forming porous structures. They can be engineered to have specific pore sizes and surface functionalities, enabling selective adsorption of various contaminants such as arsenic, lead, and pharmaceuticals (Lu et al., 2022; Svensson Grape et al., 2023).

    c. Zeolites and Clays: These materials are naturally occurring or synthetically produced aluminosilicates with a high affinity for cationic pollutants. Their microporous structures make them suitable for adsorbing heavy metals and ammonium ions from wastewater (Jaramillo-Fierro et al., 2021).

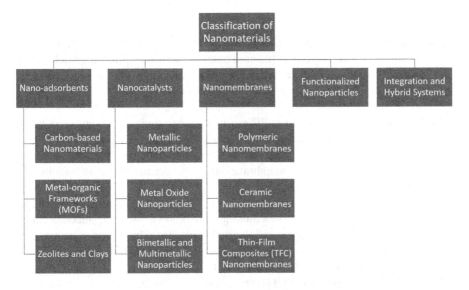

FIGURE 7.1   Various categories of nanomaterials used in wastewater treatment.

7.2.2   NANO-CATALYSTS

Nano-catalysts are nanomaterials that facilitate the degradation or transformation of pollutants through catalytic reactions. These materials often function under mild conditions, such as ambient temperature and pressure, making them energy-efficient alternatives for wastewater treatment. Key types of nano-catalysts include:

7.2.3   METALLIC NANOPARTICLES

Noble metals like platinum, palladium, and gold nanoparticles are widely used as catalysts due to their high reactivity and stability. These nanoparticles can accelerate redox reactions, aiding in the degradation of organic pollutants and reducing toxic metal ions (Ranjith et al., 2023).

a. Metal Oxide Nanoparticles: TiO_2, zinc oxide (ZnO), and iron oxide (Fe_2O_3) nanoparticles are common metal oxide catalysts. TiO_2, in particular, is extensively studied for photocatalytic applications. Under UV light, TiO_2 generates reactive oxygen species (ROS) that can break down organic contaminants into harmless byproducts (Sharma et al., 2023; Soltani et al., 2023).

b. Bimetallic and Multimetallic Nanoparticles: Combining two or more metals can enhance catalytic performance through synergistic effects. For instance, Fe/Pd bimetallic nanoparticles have improved efficiency in dechlorinating chlorinated organic compounds in wastewater (Pokkiladathu et al., 2023).

7.2.4 Nano-Membranes

Nano-membranes are nanomaterial-based membranes used for filtration and separation processes in wastewater treatment. These membranes exhibit high selectivity and permeability, enabling the efficient removal of contaminants while maintaining high water flux. Common types of nano-membranes include:

a. Polymeric Nano-membranes: These membranes typically comprise polymers like polyethersulfone (PES) and polyvinylidene fluoride (PVDF), embedded with nanomaterials such as CNTs or nanoparticles. Incorporating nanomaterials enhances membrane strength, permeability, and fouling resistance (Sahu et al., 2023a).

b. Ceramic Nano-membranes: Made from materials like alumina, zirconia, and titania, ceramic nano-membranes are known for their mechanical strength, chemical stability, and thermal resistance. These properties make them suitable for treating harsh industrial effluents containing acids, bases, and solvents (Rabia et al., 2023; Yoon et al., 2023).

c. Thin-Film Composite (TFC) Nano-membranes: TFC membranes consist of a thin selective layer, often made from polyamide, on top of a porous support layer. This configuration allows for high separation efficiency and is widely used in reverse osmosis and nanofiltration processes (Shafi et al., 2021; Shukla et al., 2023).

7.2.5 Functionalized Nanoparticles

Beyond the primary classifications, functionalized nanoparticles represent versatile nanomaterials tailored for specific applications. These nanoparticles are modified with functional groups or coatings to enhance their interaction with target pollutants. For example, magnetic nanoparticles can be functionalized with ligands that specifically bind to heavy metals, allowing for easy magnetic separation after adsorption (Azizi et al., 2023; Dai et al., 2013).

7.2.6 Integration and Hybrid Systems

The classification of nanomaterials often extends to hybrid systems where different nanomaterials are combined to leverage multiple functionalities. For instance, a hybrid system may incorporate nano-adsorbents for contaminant capture and nano-catalysts for subsequent degradation, providing a comprehensive approach to wastewater treatment (Shreve & Brennan, 2019; Thakur et al., 2023).

7.3 MECHANISMS OF NANOMATERIAL ACTION IN WASTEWATER TREATMENT

Nanomaterials have become integral to advanced wastewater treatment processes due to their unique physical and chemical properties. These materials function through various mechanisms, each contributing to effectively removing diverse contaminants. Each mechanism exploits the unique properties of nanomaterials to

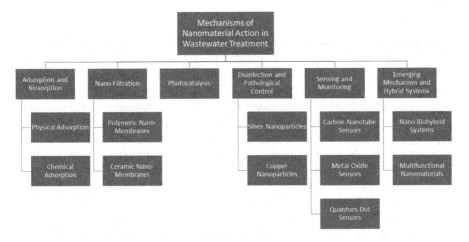

FIGURE 7.2 Various mechanisms of nanomaterial action in wastewater treatment.

enhance pollutant removal efficiency, selectivity, and cost-effectiveness. As research progresses, integrating emerging mechanisms and hybrid systems is expected to revolutionize wastewater treatment further, addressing the growing challenges of water pollution and resource management (Karnena et al., 2019; Sellamuthu et al., 2023). Figure 7.2 represents the primary mechanisms nanomaterials act in wastewater treatment, including adsorption, biosorption, nanofiltration, photocatalysis, disinfection, and sensing and monitoring.

7.3.1 ADSORPTION AND BIOSORPTION

Adsorption is one of the most widely used mechanisms for pollutant removal in wastewater treatment. Nanomaterials, such as CNTs, graphene, and MOFs, exhibit high surface areas and active sites, which facilitate the adsorption of contaminants. The adsorption process involves the interaction of pollutants with the surface of nanomaterials through physical or chemical bonds (Al-Mashhadani & Al-Mashhadani, 2023; Irshad et al., 2023).

a. Physical Adsorption: This occurs through van der Waals forces and is typically reversible. It is advantageous for removing organic pollutants and dyes due to its simplicity and effectiveness.
b. Chemical Adsorption: This involves the formation of chemical bonds between pollutants and the adsorbent surface. For example, MOFs can be engineered to have specific functional groups that chemically bond with heavy metals, enhancing selectivity and adsorption capacity. Biosorption utilizes biological materials or bio-inspired nanomaterials to adsorb contaminants. Nanoparticles coated with biomolecules or derived from biological entities (like algae or fungi) can efficiently capture heavy metals and other pollutants from wastewater. The biosorption process is highly selective and environmentally friendly.

7.3.2 Nanofiltration

Nanofiltration is a membrane filtration process that uses nanomaterial-based membranes to separate contaminants from water. These membranes have pore sizes in the nanometer range, effectively allowing them to filter out small molecules, ions, and microorganisms. The critical advantage of nanofiltration is its ability to achieve high separation efficiency while maintaining a high water flux, reducing energy consumption and operational costs (Abdelbasir & Shalan, 2019; Palani et al., 2021).

a. Polymeric Nano-membranes: Incorporating nanomaterials like CNTs or nanoparticles into polymeric membranes improves their mechanical strength, permeability, and resistance to fouling. These membranes are used in reverse osmosis and nanofiltration to remove salts, heavy metals, and organic pollutants.

b. Ceramic Nano-membranes: Made from materials such as alumina and zirconia, ceramic nano-membranes offer high thermal and chemical stability. They are suitable for treating industrial wastewater containing harsh chemicals and solvents.

7.3.3 Photocatalysis

Photocatalysis involves using nanomaterials to accelerate photochemical reactions under light irradiation, leading to the degradation of pollutants (Shivaji et al., 2023). TiO_2 is one of the most commonly used photocatalysts due to its high stability, non-toxicity, and oxidative solid power. Its mechanism involves the interaction of UV lights with TiO_2 nanoparticles, leading to electron-hole pair generation. These pairs react with water and oxygen to produce ROS, such as hydroxyl radicals and superoxide anions. ROS are highly reactive and can degrade many organic pollutants into less harmful substances, including pesticides, pharmaceuticals, and dyes (Siddique et al., 2024). Photocatalytic nanomaterials are used in various forms, such as coatings on surfaces, suspended particles, or integrated into membranes. Their ability to degrade pollutants under light makes them suitable for treating wastewater in sunlight-exposed environments (Som et al., 2020).

7.3.4 Disinfection and Pathological Control

Nanomaterials with antimicrobial properties are employed to disinfect and control pathogens in wastewater. Silver (Ag) and copper (Cu) nanoparticles are particularly effective due to their broad-spectrum antimicrobial activities. In addition, nanomaterials can be incorporated into filtration systems and coatings or used as stand-alone disinfectants to ensure the microbial safety of treated water (Bui et al., 2019; Villaseñor & Ríos, 2017).

a. Silver Nanoparticles: Ag nanoparticles release silver ions, which interact with bacterial cell membranes, proteins, and DNA, causing cell damage and death. They are effective against many bacteria, viruses, and fungi.

b. Copper Nanoparticles: Cu nanoparticles generate ROS that damage microbial cells. They are also effective in disrupting biofilms, which are communities of microorganisms that adhere to surfaces and resist conventional disinfection methods.

7.3.5 Sensing and Monitoring

Nanomaterials are also employed to detect contaminants in real-time sensing and monitoring applications. Nanosensors leverage nanomaterials' high surface area and reactivity to achieve sensitive and selective detection of pollutants. Integrating nanosensors into wastewater treatment systems allows for continuous monitoring and rapid response to contamination events, enhancing the overall efficiency and safety of the treatment process (He et al., 2023; Zhang et al., 2023).

a. CNT Sensors: CNT-based sensors can detect gases, heavy metals, and organic compounds at low concentrations due to their excellent electrical properties and surface reactivity.
b. Metal Oxide Sensors: Nanomaterials like ZnO and TiO_2 are used in sensors to detect pollutants through changes in their electrical resistance upon interaction with contaminants. These sensors are susceptible and can operate in various environmental conditions.
c. Quantum Dot Sensors: Quantum dots are semiconductor nanocrystals that exhibit size-dependent optical properties. They are used in fluorescence-based sensors to detect trace amounts of pollutants, offering high sensitivity and specificity.

7.3.6 Emerging Mechanisms and Hybrid Systems

In addition to the primary mechanisms, researchers are developing hybrid systems that combine multiple nanomaterials and mechanisms to enhance treatment efficacy. Developing such hybrid systems aims to provide comprehensive solutions for complex wastewater matrices, reducing the need for various treatment stages and lowering operational costs.

a. Nano-biohybrid Systems: These systems integrate nanomaterials with biological processes. For example, biofilm-coated nanoparticles can combine nanomaterials' high adsorption capacity with microorganisms' metabolic capabilities to degrade contaminants (Kusworo et al., 2021; Wu et al., 2017).
b. Multifunctional Nanomaterials: These materials possess multiple functionalities, such as adsorption, catalysis, and disinfection, allowing them to address a wide range of contaminants simultaneously. For instance, magnetic nanoparticles functionalized with catalytic and antimicrobial agents can adsorb, degrade, and disinfect pollutants in one step (Sadegh & Ali, 2018; Saravanan et al., 2022).

7.4 APPLICATIONS OF NANOMATERIALS IN POLLUTANT REMOVAL

Nanomaterials have shown immense potential in addressing various pollutants in wastewater due to their unique properties, such as high surface area, enhanced reactivity, and the ability to be functionalized for specific targets. Their unique properties enable them to tackle contaminants that are difficult to remove with traditional methods, thereby enhancing the overall efficacy of wastewater treatment processes. Continued advancements in nanotechnology and material science are expected to improve the performance and applicability of nanomaterials in pollutant removal (Adewuyi & Lau, 2021; Ren et al., 2023). The applications of nanomaterials in pollutant removal can be categorized into the removal of heavy metals, organic pollutants, and pathogens, as shown in Figure 7.3.

7.4.1 REMOVAL OF HEAVY METALS

Heavy metals, such as lead (Pb), mercury (Hg), cadmium (Cd), and arsenic (As), are common contaminants in industrial wastewater. These metals are highly toxic and can cause severe health issues upon exposure. Nanomaterials, particularly nano-adsorbents, have effectively adsorbed heavy metals from wastewater.

a. Carbon Nanotubes (CNTs): CNTs are highly effective in removing heavy metals due to their large surface area and ability to form strong interactions with metal ions. Functionalized CNTs with specific chemical groups can enhance selectivity for certain metals (Frayyeh et al., 2021).

b. Metal Oxide Nanoparticles: Nanoparticles such as iron oxide (Fe_3O_4) and TiO_2 are used for heavy metal removal. Fe_3O_4 nanoparticles, due to their magnetic properties, can be easily separated from water after adsorption, making them highly efficient and reusable (Lei et al., 2023; Sharma et al., 2023).

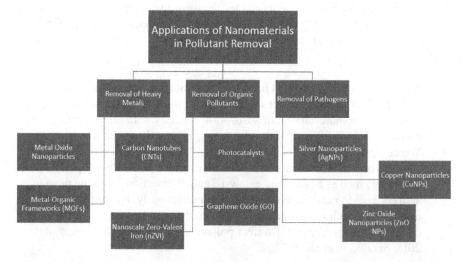

FIGURE 7.3 Different applications of nanomaterials in pollutant removal.

c. Metal-organic Frameworks (MOFs): MOFs are porous materials that can be tailored to capture specific heavy metal ions through coordination chemistry. Their tunable structures and high surface areas make them suitable for simultaneously removing multiple types of heavy metals (Karbalaee Hosseini & Tadjarodi, 2023).

7.4.2 REMOVAL OF ORGANIC POLLUTANTS

Organic pollutants, including pesticides, pharmaceuticals, and industrial chemicals, are persistent in the environment and can be challenging to remove with conventional treatment methods. Nanomaterials offer innovative solutions for the degradation and removal of these compounds.

a. Photocatalysts: TiO_2 nanoparticles are widely used as photocatalysts. Under UV light, TiO_2 generates ROS that can break down organic pollutants into less harmful substances. This process is highly effective for degrading complex organic molecules (Mehdizadeh et al., 2023).
b. Graphene Oxide (GO): GO has shown excellent adsorption capabilities for organic pollutants due to its large surface area and functional groups interacting with various organic molecules. GO can adsorb pharmaceuticals, dyes, and pesticides, making it a versatile material for water purification (Sahu et al., 2023b).
c. Nanoscale Zero-Valent Iron (nZVI): nZVI effectively reduces and dechlorinates chlorinated organic compounds. Reductive dechlorination can transform harmful pollutants like trichloroethylene (TCE) into less toxic forms (Xiao et al., 2022).

7.4.3 REMOVAL OF PATHOGENS

In wastewater, pathogens such as bacteria, viruses, and protozoa pose significant health risks. Nanomaterials with antimicrobial properties can effectively disinfect water, ensuring its safety for reuse or discharge.

a. Silver Nanoparticles (AgNPs): AgNPs are renowned for their potent antimicrobial activity. They release silver ions, which disrupt cellular processes in microbes, leading to their inactivation. AgNPs can be incorporated into filtration systems or used as surface coatings for continuous disinfection (Chhibber et al., 2017).
b. Copper Nanoparticles (CuNPs): CuNPs exhibit antimicrobial properties similar to AgNPs and are particularly effective against biofilms. They generate ROS that damage microbial cells, preventing the formation and persistence of biofilms in treatment systems (Amani & Yengejeh, 2021).
c. Zinc Oxide Nanoparticles (ZnO NPs): ZnO NPs are used for their antimicrobial properties and ability to produce ROS under light irradiation. They effectively inactivate various pathogens and can be integrated into disinfection systems (Zazouli et al., 2021).

7.5 ADVANTAGES OF NANOMATERIALS IN WASTEWATER TREATMENT

Nanomaterials have emerged as transformative agents in wastewater treatment due to their exceptional properties and capabilities. The benefits of using nanomaterials in this context are manifold, offering significant improvements over conventional treatment methods. Nanomaterials provide numerous advantages in wastewater treatment, including high efficiency, enhanced reactivity, reduced energy consumption, improved fouling resistance, multifunctionality, scalability, cost-effectiveness, environmental sustainability, and versatility. These benefits make nanomaterials a promising and essential component of modern wastewater treatment technologies, capable of addressing current and future challenges in water management. Figure 7.4 shows some of the key benefits.

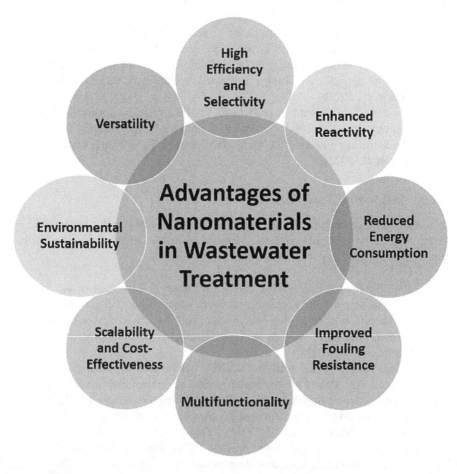

FIGURE 7.4   Advantages of nanomaterials in wastewater treatment.

7.5.1 High Efficiency and Selectivity

One of the most notable advantages of nanomaterials is their high efficiency and selectivity in removing pollutants. Due to their large surface area-to-volume ratio and high reactivity, nanomaterials can adsorb and degrade contaminants more effectively than traditional materials. For example, CNTs and MOFs can be engineered to selectively target specific pollutants, such as heavy metals and organic compounds, thereby enhancing the removal efficiency (Fu et al., 2023).

7.5.2 Enhanced Reactivity

Nanomaterials' unique electronic, optical, and catalytic properties enhance their reactivity. This is particularly beneficial in processes such as photocatalysis, where materials like TiO_2 can generate ROS under light irradiation to degrade complex organic pollutants. The high reactivity of nanomaterials also facilitates faster reaction rates, leading to quicker treatment processes (Mekonnen et al., 2021).

7.5.3 Reduced Energy Consumption

Traditional wastewater treatment methods require significant energy inputs, especially for aeration and chemical oxidation processes. Nanomaterials can reduce energy consumption by enabling more efficient treatment processes (Alali et al., 2023). For instance, nanofiltration membranes require lower pressures than reverse osmosis membranes, leading to lower energy usage. Additionally, using photocatalysts like TiO_2 can harness solar energy, reducing the need for external energy sources.

7.5.4 Improved Fouling Resistance

Fouling is a common problem in membrane-based filtration systems, reducing efficiency and increasing maintenance costs. Nanomaterials can enhance the fouling resistance of membranes. For example, incorporating AgNPs or GO into polymeric membranes can inhibit the growth of biofilms and reduce fouling, thereby extending the operational lifespan of the membranes and decreasing downtime for cleaning.

7.5.5 Multifunctionality

Nanomaterials offer multifunctional capabilities, enabling them to simultaneously perform multiple roles in wastewater treatment. For instance, magnetic nanoparticles can adsorb contaminants and be easily separated from water using a magnetic field, providing a simple and efficient pollutant removal and recovery method. Similarly, multifunctional nanomaterials can combine adsorption, catalysis, and antimicrobial actions, offering comprehensive solutions to complex wastewater treatment challenges (Miras & Alhalili, 2023).

7.5.6 SCALABILITY AND COST-EFFECTIVENESS

Advancements in nanomaterial synthesis and production techniques have made it economically possible to scale up their use in wastewater treatment. While initial costs may be higher, the long-term benefits of using nanomaterials, such as reduced energy consumption, lower maintenance requirements, and higher treatment efficiencies, can lead to overall cost savings. Additionally, regenerating and reusing specific nanomaterials, such as nano-adsorbents and catalysts, further enhances their cost-effectiveness (Lotha et al., 2024).

7.5.7 ENVIRONMENTAL SUSTAINABILITY

Nanomaterials contribute to environmental sustainability by offering greener alternatives to conventional treatment chemicals and processes. For instance, the use of photocatalysts for the degradation of pollutants reduces the need for hazardous chemicals. Furthermore, nanomaterials' high efficiency and selectivity minimize the generation of secondary waste and byproducts, reducing the environmental footprint of wastewater treatment operations (Sharma et al., 2022).

7.5.8 VERSATILITY

The versatility of nanomaterials allows them to be used in various forms and configurations, such as coatings, membranes, and dispersed particles, to address specific treatment needs. This adaptability makes nanomaterials suitable for multiple applications, from industrial wastewater treatment to potable water purification (Zieliński et al., 2022).

7.6 CHALLENGES AND CONSIDERATIONS IN NANOMATERIAL-ENABLED WASTEWATER TREATMENT

Nanomaterials hold significant promise for enhancing wastewater treatment; addressing the challenges related to environmental impact, scalability, regulation, technical feasibility, and public perception is crucial for their successful implementation. Comprehensive risk assessments, cost-effective production methods, clear regulatory frameworks, technical innovations, and transparent communication are critical to overcoming these challenges and realizing the full potential of nanomaterials in sustainable wastewater management. These challenges encompass environmental, technical, and regulatory aspects, necessitating a comprehensive approach to overcome them.

7.6.1 ENVIRONMENTAL FATE AND TOXICITY

One of the primary concerns regarding using nanomaterials in wastewater treatment is their potential environmental fate and toxicity. Nanoparticles can interact with various environmental matrices, leading to unknown ecological impacts. The release of nanomaterials into water bodies can pose risks to aquatic life due to their high reactivity and potential to accumulate in organisms. For instance, AgNPs and TiO_2 nanoparticles have been shown to exhibit toxic effects on fish and other aquatic organisms (Lee et al., 2022; Roy & Nath, 2022). To mitigate these risks, comprehensive risk assessments are essential. These assessments should evaluate the life cycle of nanomaterials, from production and usage to disposal. Developing biodegradable or environmentally benign nanomaterials can also reduce potential hazards (Dugganaboyana et al., 2023; Sadiq et al., 2023).

7.6.2 SCALABILITY AND COST

Although nanomaterials offer enhanced performance, their large-scale production and cost remain significant barriers to widespread adoption. The synthesis of high-quality nanomaterials often requires complex procedures and expensive raw materials, leading to high production costs. Moreover, scaling up laboratory processes to industrial levels without compromising the quality and consistency of nanomaterials poses technical challenges.

Addressing these issues requires advancements in nanomaterial synthesis techniques that are both cost-effective and scalable. Research into using abundant and inexpensive raw materials and developing energy-efficient production methods is crucial for making nanomaterials economically viable for large-scale wastewater treatment applications.

7.6.3 REGULATORY FRAMEWORKS

Nanomaterials' rapid development and application in wastewater treatment have outpaced the establishment of comprehensive regulatory frameworks. There is a lack of standardized guidelines for the safe use, handling, and disposal of nanomaterials. This regulatory gap can lead to inconsistent practices and potential risks to human health and the environment.

Governments and regulatory bodies must develop clear regulations addressing nanomaterials' unique properties and potential risks. These regulations should cover the entire life cycle of nanomaterials, including manufacturing, usage, monitoring, and disposal. International collaboration is essential to harmonize standards and practices across different regions (Nel et al., 2015; Schmidt, 2009).

7.6.4 TECHNICAL CHALLENGES

Several technical challenges hinder the optimal use of nanomaterials in wastewater treatment. These include:

a. Aggregation and Stability: Nanomaterials aggregate in aqueous environments, reducing their effective surface area and reactivity. Stabilizing agents or surface modifications can help maintain their dispersion and prevent aggregation (Shen et al., 2022).
b. Fouling and Deactivation: In practical applications, nanomaterials can suffer from fouling by organic matter or deactivating by wastewater chemical

agents. Developing fouling-resistant and durable nanomaterials is essential for maintaining their long-term effectiveness.

c. Integration with Existing Systems: Incorporating nanomaterials into existing wastewater treatment infrastructure requires careful consideration of compatibility and integration. Hybrid systems that combine nanomaterials with conventional treatment methods can offer synergistic benefits but require careful design and optimization.

7.6.5 PUBLIC PERCEPTION AND ACCEPTANCE

Public perception and acceptance of nanotechnology in wastewater treatment can influence its implementation. Concerns about nanomaterials' safety and ethical implications can lead to resistance from communities and stakeholders. Transparent communication about nanomaterials' benefits, risks, and safety measures is crucial for gaining public trust and acceptance.

7.7 CASE STUDIES AND REAL-WORLD APPLICATIONS OF NANOMATERIALS IN WASTEWATER TREATMENT

Nanomaterials have been successfully implemented in various real-world applications for wastewater treatment, demonstrating their effectiveness in addressing diverse contaminants and improving water quality. Some notable case studies and applications highlight the practical benefits and challenges of using nanomaterials in this field, as shown in Table 7.1.

7.8 CHALLENGES AND CONSIDERATIONS FACED DURING THIS REMEDIATION USING NANOMATERIALS

The real-world applications and case studies of nanomaterials in wastewater treatment demonstrate their potential to revolutionize water management practices. Nanomaterials offer versatile and practical solutions for removing heavy metals and organic pollutants, disinfecting water, and enhancing desalination processes (Elsaid et al., 2023; Manoj et al., 2023; New et al., 2023). Addressing the challenges will be vital to realizing the full potential of nanomaterials in creating sustainable and efficient wastewater treatment systems. While these case studies highlight the successful application of nanomaterials in wastewater treatment, several challenges and considerations must be addressed in Table 7.2.

7.9 FUTURE PROSPECTS AND RESEARCH DIRECTIONS IN NANOMATERIAL-ENABLED WASTEWATER TREATMENT

Nanomaterials have demonstrated remarkable potential in transforming wastewater treatment processes. The future prospects for nanomaterial-enabled wastewater treatment are promising, with numerous research directions offering pathways to overcome current limitations and enhance these technologies' efficiency, safety, and

TABLE 7.1

Different Case Studies with its Treating Nanomaterial and Pollutant Addressed with Key Benefits and Challenges

Sr. No.	Case study/Application	Location	Nanomaterial used	Pollutant Addressed	Key Benefits	Challenges	Reference
1.	Arsenic Removal	Bangladesh	Iron oxide nanoparticles	Arsenic	Reduced arsenic levels to safe limits	Environmental fate and toxicity	Yang et al. (2019)
2.	Industrial Wastewater Treatment	China	Titanium dioxide (TiO_2)	Dyes and organic pollutants	Degradation of organic pollutants under UV light	Cost and scalability	Cervantes-Avilés et al. (2022)
3.	Municipal Wastewater Treatment	United States	Graphene oxide (GO)	Pharmaceuticals, personal care products	Enhanced permeability and fouling resistance	Integration with existing systems	Subtil et al. (2022)
4.	Pathogen Removal	South Africa	Silver nanoparticles (AgNPs)	Pathogens (bacteria, viruses)	Effective disinfection and microbial safety	Public acceptance	Palani et al. (2023)
5.	Heavy Metal Removal	India	Magnetic iron oxide (Fe_3O_4)	Lead (Pb), cadmium (Cd)	Selective heavy metal removal, easy separation using magnetic field	Scalability and cost	Kulpa-Koterwa et al. (2022)
6.	Organic Pollutant Degradation	Europe	Palladium (Pd) and iron (Fe) bimetallic nanoparticles	Chlorinated organic compounds	Efficient dechlorination and pollutant breakdown	Regulatory compliance	Ulucan-Altuntas & Debik (2020)
7.	Desalination and Water Reuse	Middle East	Carbon nanotubes (CNTs)	Salts	Improved desalination efficiency	Scalability and cost	Chowdhury et al. (2018)

TABLE 7.2

Different Challenges During Remediation Using Nanomaterials with its Details

S. No.	Challenges	Details
1.	Scalability	Scaling up production while maintaining quality and consistency is challenging.
2.	Cost	High synthesis costs of high-quality nanomaterials can hinder widespread adoption.
3.	Regulatory compliance	There is a need for clear regulatory frameworks for the safe use, handling, and disposal of nanomaterials.
4.	Public acceptance	Gaining trust and acceptance through transparent communication about benefits and risks.

sustainability. Developing novel nanomaterials, advancing synthesis techniques, integrating with existing systems, ensuring environmental and health protection, and addressing emerging contaminants is critical for future research. Additionally, real-time monitoring, cost-effective regeneration, and public engagement will be crucial for successfully implementing and accepting nanomaterial-based wastewater treatment solutions. Through continued innovation and collaboration, nanomaterials can play a pivotal role in addressing global water challenges and ensuring the availability of clean water for future generations. However, there are numerous avenues for future research and development to harness their capabilities further and address existing limitations. This section outlines the future prospects and research directions in nanomaterial-enabled wastewater treatment.

7.9.1 Development of Novel Nanomaterials

Creating new nanomaterials with enhanced properties is critical for future research. Innovations in material science can lead to the development of nanomaterials that offer even higher adsorption capacities, selectivity, and stability.

a. Biocompatible and Biodegradable Nanomaterials: Developing biocompatible and biodegradable nanomaterials can mitigate environmental concerns associated with their use. Such materials would break down into non-toxic components after their intended use, reducing the risk of environmental contamination (Kapoor & Dhawan, 2013).
b. Hybrid Nanomaterials: Combining different types of nanomaterials to create hybrids can offer multifunctional properties. For example, integrating magnetic nanoparticles with catalytic materials can facilitate pollutant removal and easy recovery of the nanomaterials using a magnetic field (Zeng & Xia, 2012).
c. Smart Nanomaterials: These materials can respond to environmental stimuli, such as pH, temperature, or light. Innovative nanomaterials can change their properties or activate specific functions in response to environmental changes, leading to more efficient and targeted pollutant removal (Abu-Dief & Mohamed, 2020).

7.9.2   ADVANCED SYNTHESIS TECHNIQUES

Improving synthesis methods to produce nanomaterials at lower costs and higher scales is essential for their broader adoption in wastewater treatment.

a. Green Synthesis: Developing environmentally friendly synthesis methods using non-toxic solvents, biological agents, and sustainable raw materials can make nanomaterials more sustainable (Singh et al., 2020).
b. Scalable Production: Techniques that enable the scalable production of nanomaterials without compromising their quality and properties are crucial. This includes chemical vapor deposition, sol-gel processes, and hydrothermal synthesis methods (Mishra & Ahmaruzzaman, 2022).
c. Controlled Morphologies: Advances in controlling the size, shape, and surface functionalities of nanomaterials can enhance their performance. For instance, tuning the pore size in MOFs can improve their selectivity for specific pollutants (Zamel et al., 2023).

7.9.3   INTEGRATION WITH EXISTING TREATMENT SYSTEMS

Integrating nanomaterials with conventional wastewater treatment systems can leverage the strengths of both approaches, leading to more efficient and comprehensive treatment solutions.

a. Hybrid Treatment Systems: Combining nanomaterials with existing physical, chemical, and biological treatment methods can provide synergistic effects. For example, integrating nanomaterial-based adsorbents with biological treatment processes can enhance the overall removal efficiency of complex contaminants (Bhuvanendran & Bhuvaneshwari, 2023).
b. Retrofitting Existing Infrastructure: Research into how nanomaterials can be incorporated into existing treatment plants without requiring extensive modifications is vital. This could involve developing nanomaterial-coated membranes or adding nanomaterials to existing filtration units (Lusiana et al., 2020).

7.9.4   ENVIRONMENTAL AND HEALTH SAFETY

Ensuring nanomaterials' environmental and health safety throughout their lifecycle is critical for sustainable wastewater treatment.

a. Life Cycle Assessments (LCAs): Conducting comprehensive LCAs can help understand the environmental impact of nanomaterials from production to disposal. This includes assessing potential risks to human health and ecosystems (Teodosiu et al., 2016).
b. Toxicological Studies: More research is needed to understand the toxicity of nanomaterials, especially their long-term effects on aquatic life and human health. This includes studying the mechanisms of toxicity and identifying safe concentration limits (Grigoras, 2020).

c. Regulatory Frameworks: Developing clear and consistent regulatory guidelines for using, handling, and disposing of nanomaterials is essential. This requires collaboration between scientists, policymakers, and industry stakeholders (Zimmer et al., 2014).

7.9.5 REAL-TIME MONITORING AND CONTROL

Advanced sensing and monitoring technologies can enhance the management and control of wastewater treatment processes.

a. Nanosensors: The development of highly sensitive and selective nanosensors for real-time monitoring of contaminants can provide immediate feedback and allow for dynamic adjustments in treatment processes. These sensors can detect trace levels of pollutants, enabling more precise control over the treatment process (Kumar & Guleria, 2020).

b. Internet of Things (IoT) Integration: Integrating nanomaterial-based sensors with IoT platforms can enable remote monitoring and control of wastewater treatment systems. This can lead to more efficient operation and maintenance of treatment plants (Hoa et al., 2023).

7.9.6 ADDRESSING EMERGING CONTAMINANTS

Emerging contaminants, such as pharmaceuticals, personal care products, and microplastics, present new challenges for wastewater treatment. Nanomaterials offer promising solutions for addressing these pollutants.

a. Pharmaceuticals and Personal Care Products: Research into using nanomaterials to degrade and remove these persistent pollutants is critical. Photocatalytic nanomaterials, for example, can break down complex organic molecules into less harmful byproducts (Freyria et al., 2018).

b. Microplastics: Developing nanomaterial-based filtration systems that can capture microplastics from wastewater before they enter natural water bodies is an emerging area of research. Magnetic nanoparticles, for instance, can be functionalized to attract and remove microplastics (Arat, 2024).

7.9.7 COST-EFFECTIVE REGENERATION AND REUSE

The ability to regenerate and reuse nanomaterials without significant performance loss is crucial for their economic feasibility.

a. Regeneration Techniques: Developing efficient methods for regenerating spent nanomaterials can reduce operational costs. This includes thermal treatment, chemical regeneration, and electrochemical processes.

b. Longevity and Durability: Researching ways to enhance the longevity and durability of nanomaterials can reduce the frequency of replacement and improve overall cost-effectiveness. This involves enhancing resistance to fouling, chemical degradation, and mechanical wear.

7.9.8 Public Engagement and Education

Public perception and acceptance play a significant role in deploying nanomaterial-based technologies. Engaging with the public and educating them about the benefits and safety of these technologies is essential.

a. Community Outreach: Initiatives to inform communities about the advantages of nanomaterials in wastewater treatment and address any concerns that can build public trust. This includes transparent communication about the risks and benefits.
b. Educational Programs: Integrating nanotechnology topics into educational curricula can raise awareness and understanding among future scientists, engineers, and policymakers.

7.10 CONCLUSIONS

Nanomaterials have significant potential to revolutionize wastewater treatment, providing innovative solutions to longstanding environmental challenges. Their unique properties, such as high surface area, enhanced reactivity, and functionalization ability, enable efficient removal of pollutants like heavy metals, organic contaminants, and pathogens. Integrating nanomaterials into treatment processes has improved efficiency, selectivity, and cost-effectiveness, offering a promising alternative to traditional methods. However, challenges remain, including environmental and health safety concerns, scalability issues, high production costs, and the need for robust regulatory frameworks. Addressing these issues through advanced synthesis techniques, comprehensive risk assessments, and the development of environmentally benign nanomaterials is crucial for sustainable use. Future prospects for nanomaterial-enabled wastewater treatment are promising, with potential advancements in novel nanomaterials, hybrid systems, and intelligent technologies. These will enhance versatility and effectiveness, making nanomaterials suitable for broader applications. Integrating real-time monitoring systems and IoT platforms can also optimize treatment processes, ensuring better water quality control and management. In conclusion, nanomaterials offer transformative potential for wastewater treatment, paving the way for sustainable and efficient water management practices. Continued innovation and interdisciplinary collaboration are essential to overcome challenges and fully realize the benefits of nanomaterials in providing clean and safe water for all.

REFERENCES

Abdelbasir, S. M.; Shalan, A. E. An Overview of Nanomaterials for Industrial Wastewater Treatment. *Korean Journal of Chemical Engineering* **2019**, *36* (8), pp. 1209–1225. https://doi.org/10.1007/S11814-019-0306-Y.

Abu-Dief, A. M.; Mohamed, W. S. Development of Nanomaterials as Photo Catalysts for Environmental Applications. *Current Catalysis* **2020**, *9* (2), pp. 128–137. https://doi.org/10.2174/2211544709999201123193710.

Adewuyi, A.; Lau, W. J. *Nanomaterial Development and Its Applications for Emerging Pollutant Removal in Water*; Elsevier, **2021**. https://doi.org/10.1016/B978-0-12-821506-7.00003-X.

Alali, Y.; Harrou, F.; Sun, Y. Unlocking the Potential of Wastewater Treatment: Machine Learning Based Energy Consumption Prediction. *Water* **2023**, *15* (13), p. 2349. https://doi.org/10.3390/W15132349.

Al-Mashhadani, E. S. M.; Al-Mashhadani, M. K. H. Utilization of Chlorella Vulgaris after the Extraction Process in Wastewater Treatment as a Biosorption Material for Ciprofloxacin Removal. *Journal of Ecological Engineering* **2023**, *24* (4), pp. 1–15. https://doi.org/10.12911/22998993/159336.

Amani, M. G.; Yengejeh, R. J. Comparison of Escherichia Coli and Klebsiella Removal Efficiency in Aquatic Environments Using Silver and Copper Nanoparticles. *Journal of Health Sciences and Surveillance System* **2021**, *9* (2), pp. 72–80. https://doi.org/10.30476/JHSSS.2021.88646.1151.

Arat, S. A. A Review of microplastics in Wastewater Treatment Plants in Türkiye: Characteristics, Removal Efficiency, Mitigation Strategies for Microplastic Pollution and Future Perspective. *Water Science and Technology* **2024**, *89* (7), pp. 1771–1786. https://doi.org/10.2166/WST.2024.082/1381728/WST2024082.PDF.

Azizi, shahab; Jafarbeigi, E.; Salimi, F. Removal of Heavy Metals and Dye from Water and Wastewater Using Nanofiltration Membranes of Polyethersulfone Modified with Functionalized Iron-Silica Nanoparticles. *Energy Sources, Part A: Recovery, Utilization, and Environmental Effects* **2023**, *45* (3), pp. 6856–6868. https://doi.org/10.1080/15567036.2023.2219222.

Bandehali, S.; Parvizian, F.; Moghadassi, A.; Hosseini, S. M. Nanomaterials for the Efficient Abatement of Wastewater Contaminants by Means of Reverse Osmosis and Nanofiltration. *Nanomaterials for the Detection and Removal of Wastewater Pollutants* **2020**, pp. 111–144. https://doi.org/10.1016/B978-0-12-818489-9.00005-0.

Bhuvanendran, R. K.; Bhuvaneshwari, S. Hybrid Electrocoagulation Reactor for Dairy Wastewater Treatment and Methodology for Sludge Reusability for the Development of Vermicompost. *Environmental Science and Pollution Research* **2023**, *30* (39), pp. 90960–90979. https://doi.org/10.1007/S11356-023-28805-1/METRICS.

Bui, X.-T., Chiemchaisri, C., Fujioka, T., Varjani, S., Eds.; *Water and Wastewater Treatment Technologies; Energy, Environment, and Sustainability*; Springer Singapore: Singapore, **2019**. https://doi.org/10.1007/978-981-13-3259-3.

Canbaz, G. T.; Çakmak, N. K.; Eroğlu, A.; Açıkel, Ü. Removal of Acid Orange 74 from Wastewater with TiO$_2$ Nanoparticle. *International Advanced Researches and Engineering Journal* **2019**, *3* (1), pp. 75–80.

Cervantes-Avilés, P.; Saber, A. N.; Mora, A.; Mahlknecht, J.; Cuevas-Rodríguez, G. Correction to: Influence of Wastewater Type in the Effects Caused by Titanium Dioxide Nanoparticles in the Removal of Macronutrients by Activated Sludge (Environmental Science and Pollution Research, (2022), 29, 6, (8746-8757), 10.1007/S11356-021-16221-2). *Environmental Science and Pollution Research* **2022**, *29* (6), p. 8758. https://doi.org/10.1007/S11356-021-17260-5/METRICS.

Chadha, U.; Selvaraj, S. K.; Vishak Thanu, S.; Cholapadath, V.; Abraham, A. M.; Zaiyan, M.; Manikandan, M.; Paramasivam, V. A Review of the Function of Using Carbon Nanomaterials in Membrane Filtration for Contaminant Removal from Wastewater. *Materials Research Express* **2022**. https://doi.org/10.1088/2053-1591/ac48b8.

Chandran, D. G.; Muruganandam, L.; Biswas, R. A Review on Adsorption of Heavy Metals from Wastewater Using Carbon Nanotube and Graphene-Based Nanomaterials. *Environmental Science and Pollution Research* **2023**, *30* (51), pp. 110010–110046. https://doi.org/10.1007/S11356-023-30192-6/METRICS.

Chhibber, S.; Gondil, V. S.; Sharma, S.; Kumar, M.; Wangoo, N.; Sharma, R. K. A Novel Approach for Combating Klebsiella Pneumoniae Biofilm Using Histidine Functionalized Silver Nanoparticles. *Frontiers in Microbiology* **2017**, *8* (JUN), p. 261401. https://doi.org/10.3389/FMICB.2017.01104/BIBTEX.

Chowdhury, Z. Z.; Sagadevan, S.; Johan, R. Bin; Shah, S. T.; Adebesi, A.; Md, S. I.; Rafique, R. F. A Review on Electrochemically Modified Carbon Nanotubes (CNTs) Membrane for Desalination and Purification of Water. *Materials Research Express* **2018**, *5* (10), p. 102001. https://doi.org/10.1088/2053-1591/AADA65.

Dai, M.; Li, J.; Kang, B.; Ren, C.; Chang, S.; Dai, Y. Magnetic Nanoparticle Decorated Multi-Walled Carbon Nanotubes for Removing Copper Ammonia Complex from Water. *Journal of Nanoscience Nanotechnology* **2013**, *13* (3), pp. 1927–1930. https://doi.org/10.1166/JNN.2013.7105.

Dugganaboyana, G. K.; Kumar Mukunda, C.; Jain, A.; Kantharaju, R. M.; Nithya, R. R.; Ninganna, D.; Ahalliya, R. M.; Shati, A. A.; Alfaifi, M. Y.; Elbehairi, S. E. I.; Silina, E.; Stupin, V.; Velliyur Kanniappan, G.; Achar, R. R.; Shivamallu, C.; Kollur, S. P. Environmentally Benign Silver Bio-Nanomaterials as Potent Antioxidant, Antibacterial, and Antidiabetic Agents: Green Synthesis Using Salacia Oblonga Root Extract. *Frontiers in Chemistry* **2023**, *11*.

Elsaid, K.; Olabi, A. G.; Abdel-Wahab, A.; Elkamel, A.; Alami, A. H.; Inayat, A.; Chae, K. J.; Abdelkareem, M. A. Membrane Processes for Environmental Remediation of Nanomaterials: Potentials and Challenges. *Science of the Total Environment* **2023**, *879*, p. 162569. https://doi.org/10.1016/J.SCITOTENV.2023.162569.

El-Sayyad, G. S.; Abd Elkodous, M.; El-Khawaga, A. M.; Elsayed, M. A.; El-Batal, A. I.; Gobara, M. Merits of Photocatalytic and Antimicrobial Applications of Gamma-Irradiated CoxNi1−xFe2O4/SiO2/TiO2; x = 0.9 Nanocomposite for Pyridine Removal and Pathogenic Bacteria/Fungi Disinfection: Implication for Wastewater Treatment. *RSC Advances* **2020**, *10* (9), pp. 5241–5259. https://doi.org/10.1039/C9RA10505K.

Frayyeh, W. M.; Mohammed, Z. B.; Mahmood, A. K. Removal of Heavy Metal Ions from Wastewater by Carbon Nanotubes (CNTs). *Engineering and Technology Journal* **2021**, *39* (12), pp. 1939–1944. https://doi.org/10.30684/ETJ.V39I12.789.

Freyria, F. S.; Geobaldo, F.; Bonelli, B. Nanomaterials for the Abatement of Pharmaceuticals and Personal Care Products from Wastewater. *Applied Sciences* **2018**, *8* (2), p. 170. https://doi.org/10.3390/APP8020170.

Fu, C.; Li, D.; Zhang, J.; Guo, W.; Yang, H.; Zhao, B.; Chen, Z.; Fu, X.; Liang, Z.; Jiang, L. Vertical 3D Printed Pd/TiO2 Arrays for High Efficiency Photo-Assisted Catalytic Water Treatment. *Chemical Research in Chinese Universities* **2023**, *39* (6), pp. 891–901. https://doi.org/10.1007/S40242-023-3182-2/METRICS.

Grigoras, A. G. Pseudomonas Species for Environmental Cleaning of Toxic Heavy Metals; **2020**. https://doi.org/10.1007/978-3-030-48985-4_1.

He, F.; Zhu, M.; Fan, J.; Ma, E.; Zhai, S.; Zhao, H. Automated Drone-Delivery Solar-Driven Onsite Wastewater Smart Monitoring and Treatment System. *Advanced Science* **2023**, *10* (24), p. 2302935. https://doi.org/10.1002/ADVS.202302935.

Hoa, N. T. N.; Dat, L. V.; Phuoc, P. H.; Manh, H. Van; Viet, N. N.; Hieu, N. Van. Integrating Multi-Sensor Chip and IoT Technology for Precise Monitoring of CO and NO2 Gas Levels in Garage Environment. In Proceedings of the 7th International Conference on Engineering Mechanics and Automation; Publishing House for Science and Technology, *Vietnam Academy of Science and Technology (Publications)*, **2023**. https://doi.org/10.15625/VAP.2023.0118.

Irshad, M. A.; Nawaz, R.; Wojciechowska, E.; Mohsin, M.; Nawrot, N.; Nasim, I.; Hussain, F. Application of Nanomaterials for Cadmium Adsorption for Sustainable Treatment of Wastewater: A Review. *Water, Air, & Soil Pollution* **2023**, *234* (1), pp. 1–17. https://doi.org/10.1007/S11270-023-06064-7/METRICS.

Jaramillo-Fierro, X.; González, S.; Montesdeoca-Mendoza, F.; Medina, F. Structuring of Zntio3 /Tio2 Adsorbents for the Removal of Methylene Blue, Using Zeolite Precursor Clays as Natural Additives. *Nanomaterials* **2021**, *11* (4), p. 898. https://doi.org/10.3390/ NANO11040898/S1.

Kamyab, H.; Chelliapan, S.; Hayder, G.; Yusuf, M.; Taheri, M. M.; Rezania, S.; Hasan, M.; Yadav, K. K.; Khorami, M.; Farajnezhad, M.; Nouri, J. Exploring the Potential of Metal and Metal Oxide Nanomaterials for Sustainable Water and Wastewater Treatment: A Review of Their Antimicrobial Properties. *Chemosphere* **2023**, *335*, p. 139103. https:// doi.org/10.1016/J.CHEMOSPHERE.2023.139103.

Kapoor, D. N.; Dhawan, S. *Biocompatible and Biodegradable Nanomaterials;* ASME Press, **2013**. https://doi.org/10.1115/1.860113_CH5.

Karbalaee Hosseini, A.; Tadjarodi, A. Novel Zn Metal–Organic Framework with the Thiazole Sites for Fast and Efficient Removal of Heavy Metal Ions from Water. *Scientific Reports* **2023**, *13* (1), pp. 1–9. https://doi.org/10.1038/s41598-023-38523-w.

Karnena, M. K.; Konni, M.; Saritha, V. Nano-Catalysis Process for Treatment of Industrial Wastewater. *Handbook of Research on Emerging Developments and Environmental Impacts of Ecological Chemistry* **2019**, pp. 229–251. https://doi.org/10.4018/978-1-7998-1241-8. CH011.

Kulpa-Koterwa, A.; Ryl, J.; Górnicka, K.; Niedziałkowski, P. New Nanoadsorbent Based on Magnetic Iron Oxide Containing 1,4,7,10-Tetraazacyclododecane in Outer Chain (Fe3O4@SiO2-Cyclen) for Adsorption and Removal of Selected Heavy Metal Ions Cd^{2+}, Pb^{2+}, Cu^{2+}. *Journal of Molecular Liquids* **2022**, *368*, p. 120710. https://doi. org/10.1016/J.MOLLIQ.2022.120710.

Kumar, V.; Guleria, P. Application of DNA-Nanosensor for Environmental Monitoring: Recent Advances and Perspectives. *Current Pollution Reports* **2020**, pp. 1–21. https://doi. org/10.1007/S40726-020-00165-1/FIGURES/10.

Kusworo, T. D.; Aryanti, N.; Utomo, D. P.; Nurmala, E. Performance Evaluation of PES-ZnO Nanohybrid Using a Combination of UV Irradiation and Cross-Linking for Wastewater Treatment of the Rubber Industry to Clean Water. *Journal of Membrane Science and Research* **2021**, *7* (1), pp. 4–13. https://doi.org/10.22079/JMSR.2020.120490.1334.

Lazarenko, N. S.; Golovakhin, V. V.; Shestakov, A. A.; Lapekin, N. I.; Bannov, A. G. Recent Advances on Membranes for Water Purification Based on Carbon Nanomaterials. *Membranes (Basel)* **2022**, *12* (10), p. 915. https://doi.org/10.3390/MEMBRANES12100915.

Lee, Y. L.; Shih, Y. S.; Chen, Z. Y.; Cheng, F. Y.; Lu, J. Y.; Wu, Y. H.; Wang, Y. J. Toxic Effects and Mechanisms of Silver and Zinc Oxide Nanoparticles on Zebrafish Embryos in Aquatic Ecosystems. *Nanomaterials* **2022**, *12* (4), p. 717. https://doi.org/10.3390/ NANO12040717/S1.

Lei, T.; Jiang, X.; Zhou, Y.; Chen, H.; Bai, H.; Wang, S.; Yang, X. A Multifunctional Adsorbent Based on 2,3-Dimercaptosuccinic Acid/Dopamine-Modified Magnetic Iron Oxide Nanoparticles for the Removal of Heavy-Metal Ions. *Journal of Colloid and Interface Science* **2023**, *636*, pp. 153–166. https://doi.org/10.1016/J.JCIS.2023.01.011.

Lemessa, G.; Chebude, Y.; Alemayehu, E. Synthesis of Cellulose-Based Nano Composites as Sustainable Adsorbents for Adsorption Removal of Heavy Metals from Wastewater. *A Review. African Journal of Water Conservation and Sustainability* **2022**, *10* (1), pp. 1–12.

Lotha, T. N; Sorhie, V.; Bharali, P.; Jamir, L. Advancement in Sustainable Wastewater Treatment: A Multifaceted Approach to Textile Dye Removal through Physical, Biological and Chemical Techniques. *ChemistrySelect* **2024**, *9* (11), p. e202304093. https://doi.org/10.1002/SLCT.202304093.

Lu, Y.; Rakshagan, V.; Shoukat, S.; Mahmoud, M. Z.; Pustokhina, I.; Salah Al-Shati, A.; Ibrahim Namazi, N.; Alshehri, S.; AboRas, K. M.; Abourehab, M. A. S. Molecular Separation and Computational Simulation of Contaminant Removal from Wastewater Using Zirconium UiO-66-(CO$_2$H)$_2$ Metal–Organic Framework. *Journal of Molecular Liquids* **2022**, *365*, p. 120178. https://doi.org/10.1016/J.MOLLIQ.2022.120178.

Lusiana, N.; Rahadi, B.; Anggita, Y. Determination Pollution Load Capacity of Ngrowo River as Wastewater Receiver from Hospital Activities. *IOP Conference Series: Earth and Environmental Science* **2020**, *475* (1), p. 012067. https://doi.org/10.1088/1755-1315/475/1/012067.

Manoj, D.; Saravanan, R.; Raji, A.; Thangamani, A. Carbon-Based Microelectrodes for Environmental Remediation: Progress, Challenges and Opportunities. *Carbon Letters* **2023**, *33* (6), pp. 1485–1493. https://doi.org/10.1007/S42823-023-00587-Z/METRICS.

Mehdizadeh, P.; Jamdar, M.; Mahdi, M. A.; Abdulsahib, W. K.; Jasim, L. S.; Raheleh Yousefi, S.; Salavati-Niasari, M. Rapid Microwave Fabrication of New Nanocomposites Based on Tb-Co-O Nanostructures and Their Application as Photocatalysts under UV/Visible Light for Removal of Organic Pollutants in Water. *Arabian Journal of Chemistry* **2023**, *16* (4), p. 104579. https://doi.org/10.1016/J.ARABJC.2023.104579.

Mekonnen, T. B.; Mengesha, A. T.; Dube, H. H. Photocatalytic Degradation of Organic Pollutants: The Case of Conductive Polymer Supported Titanium Dioxide (TiO$_2$) Nanoparticles: A Review. *Nanoscience and Nanometrology* **2021**, *7* (1), pp. 1–13. https://doi.org/10.11648/J.NSNM.20210701.11.

Miras, N.; Alhalili, Z. Metal Oxides Nanoparticles: General Structural Description, Chemical, Physical, and Biological Synthesis Methods, Role in Pesticides and Heavy Metal Removal through Wastewater Treatment. *Molecule* **2023**, *28* (7), p. 3086. https://doi.org/10.3390/MOLECULES28073086.

Mishra, S. R.; Ahmaruzzaman, M. Tin Oxide Based Nanostructured Materials: Synthesis and Potential Applications. *Nanoscale* **2022**, *14* (5), pp. 1566–1605. https://doi.org/10.1039/D1NR07040A.

Nel, A. E.; Parak, W. J.; Chan, W. C. W.; Xia, T.; Hersam, M. C.; Brinker, C. J.; Zink, J. I.; Pinkerton, K. E.; Baer, D. R.; Weiss, P. S. Where Are We Heading in Nanotechnology Environmental Health and Safety and Materials Characterization? *ACS Nano* **2015**, *9* (6), pp. 5627–5630. https://doi.org/10.1021/ACSNANO.5B03496/ASSET/IMAGES/LARGE/NN-2015-03496Q_0011.JPEG.

New, W. X.; Ogbezode, J. E.; Gani, P. Nanoparticles in Soil Remediation: Challenges and Opportunities. *Industrial and Domestic Waste Management* **2023**, *3* (2), pp. 127–140. https://doi.org/10.53623/IDWM.V3I2.357.

Oun, A. A.; Shankar, S.; Rhim, J. W. Multifunctional Nanocellulose/Metal and Metal Oxide Nanoparticle Hybrid Nanomaterials. *Critical Reviews in Food Science and Nutrition* **2020**, *60* (3), pp. 435–460. https://doi.org/10.1080/10408398.2018.1536966.

Palani, G.; L, A. latha; K, K.; Lakkaboyana, S. K.; Hanafiah, M. M.; Kumar, V.; Marella, R. K. Advanced Nanomaterial's for Industrial Wastewater Treatment – A Review. **2021**. https://doi.org/10.20944/PREPRINTS202102.0337.V1.

Palani, G.; Trilaksana, H.; Sujatha, R. M.; Kannan, K.; Rajendran, S.; Korniejenko, K.; Nykiel, M.; Uthayakumar, M. Silver Nanoparticles for Waste Water Management. *Molecules* **2023**, *28* (8), p. 3520. https://doi.org/10.3390/MOLECULES28083520.

Palanivel, P. D.; Hariharan, P.; Agilandeswari, K. Investigation of Biosorption Properties of Water Hyacinth Root in Textile Effluent and Synthetic Wastewater Treatment. *Journal of Water Chemistry and Technology* **2023**, *45* (4), pp. 343–357. https://doi.org/10.3103/S1063455X23040112.

Pokkiladathu, H.; Farissi, S.; Muthukumar, A.; Muthuchamy, M. Removal of a Contaminant of Emerging Concern by Heterogeneous Catalytic Ozonation Process with a Novel Nano Bimetallic Catalyst Embedded on Activated Carbon. *Ozone: Science & Engineering* **2023**, *45* (4), pp. 361–373. https://doi.org/10.1080/01919512.2022.2114419.

Rabia, A. R.; Adam, T.; Gopinath, S. C. B. Process Optimization and Adsorption Isotherms Investigation in the Removal of Pb2+ from Wastewater Using a Nano-Activated Alumina Membrane (NAAM). *Biomass Conversion and Biorefinery* **2023**, *13* (15), pp. 13621–13632. https://doi.org/10.1007/S13399-022-03037-4/METRICS.

Ranjith, R.; Vignesh, S.; Balachandar, R.; Suganthi, S.; Raj, V.; Ramasundaram, S.; Kalyana Sundar, J.; Shkir, M.; Oh, T. H. Construction of Novel G-C3N4 Coupled Efficient Bi2O3 Nanoparticles for Improved Z-Scheme Photocatalytic Removal of Environmental Wastewater Contaminant: Insight Mechanism. *Journal of Environmental Management* **2023**, *330*, p. 117134. https://doi.org/10.1016/J.JENVMAN.2022.117134.

Ren, C.; Bai, R.; Chen, W.; Li, J.; Zhou, X.; Tian, X.; Zhao, F. Advances in Nanomaterial-Microbe Coupling System for Removal of Emerging Contaminants. *Chemical Research in Chinese Universities* **2023**, *39* (3), pp. 389–394. https://doi.org/10.1007/S40242-023-3053-X/METRICS.

Roy, P.; Nath, D. Toxic Effects of Metal Nanoparticles on Fish Models: A Review. *Uttar Pradesh Journal of Zoology* **2022**, *43* (23), pp. 25–36. https://doi.org/10.56557/UPJOZ/2022/V43I233247.

Sadegh, H.; Ali, G. A. M. Potential Applications of Nanomaterials in Wastewater Treatment; **2018**. https://doi.org/10.4018/978-1-5225-5754-8.CH004.

Sadiq, M. U.; Shah, A.; Haleem, A.; Shah, S. M.; Shah, I. Eucalyptus Globulus Mediated Green Synthesis of Environmentally Benign Metal Based Nanostructures: A Review. *Nanomaterials* **2023**, *13* (13), p. 2019. https://doi.org/10.3390/NANO13132019.

Sahu, A.; Dosi, R.; Kwiatkowski, C.; Schmal, S.; Poler, J. C. Advanced Polymeric Nanocomposite Membranes for Water and Wastewater Treatment: A Comprehensive Review. *Polymers (Basel)* **2023a**, *15* (3), p. 540. https://doi.org/10.3390/POLYM15030540.

Sahu, P. S.; Verma, R. P.; Tewari, C.; Sahoo, N. G.; Saha, B. Facile Fabrication and Application of Highly Efficient Reduced Graphene Oxide (RGO)-Wrapped 3D Foam for the Removal of Organic and Inorganic Water Pollutants. *Environmental Science and Pollution Research* **2023b**, *30* (40), pp. 93054–93069. https://doi.org/10.1007/S11356-023-28976-X/METRICS.

Saravanan, A.; Kumar, P. S.; Hemavathy, R. V.; Jeevanantham, S.; Jawahar, M. J.; Neshaanthini, J. P.; Saravanan, R. A Review on Synthesis Methods and Recent Applications of Nanomaterial in Wastewater Treatment: Challenges and Future Perspectives. *Chemosphere* **2022**, *307*, p. 135713. https://doi.org/10.1016/J.CHEMOSPHERE.2022.135713.

Schmidt, C. W. Nanotechnology-Related Environment, Health, and Safety Research: Examining the National Strategy. *Environmental Health Perspectives* **2009**, *117* (4), pp. A158–A161. https://doi.org/10.1289/EHP.117-A158/ASSET/79602B72-8EDA-4D92-A8BB-7F483FB3FBBC/ASSETS/GRAPHIC/EHP-117-A158F3.JPG.

Sellamuthu, B.; Ramachandran, A. S. K.; Kumar, G. Extraction of Heavy Metal Ion from Industrial Wastewater by Using Bio-Nanomaterial Composite. *AIP Conference Proceedings* **2023**, *2766* (1), p. 020087. https://doi.org/10.1063/5.0140581/2894928.

Shafi, Q. I.; Ihsan, H.; Hao, Y.; Wu, X.; Ullah, N.; Younas, M.; He, B.; Rezakazemi, M. Multi-Ionic Electrolytes and E.Coli Removal from Wastewater Using Chitosan-Based in-Situ Mediated Thin Film Composite Nanofiltration Membrane. *Journal of Environmental Management* **2021**, *294*, p. 112996. https://doi.org/10.1016/J.JENVMAN.2021.112996.

Sharma, A.; Mittal, R.; Sharma, P.; Pal, K.; Mona, S. Sustainable Approach for Adsorptive Removal of Cationic and Anionic Dyes by Titanium Oxide Nanoparticles Synthesized Biogenically Using Algal Extract of Spirulina. *Nanotechnology* **2023**, *34* (48), p. 485301. https://doi.org/10.1088/1361-6528/ACF37E.

Sharma, M.; Yadav, A.; Mandal, M. K.; Pandey, S.; Pal, S.; Chaudhuri, H.; Chakrabarti, S.; Dubey, K. K. *Wastewater Treatment and Sludge Management Strategies for Environmental Sustainability;* Elsevier, **2022**; Vol. 2. https://doi.org/10.1016/B978-0-12-821664-4.00027-3.

Sharma, R.; Garg, R.; Bali, M.; Eddy, N. O. Biogenic Synthesis of Iron Oxide Nanoparticles Using Leaf Extract of Spilanthes Acmella: Antioxidation Potential and Adsorptive Removal of Heavy Metal Ions. *Environmental Monitoring and Assessment* **2023**, *195* (11), pp. 1–17. https://doi.org/10.1007/S10661-023-11860-Z/METRICS.

Shen, C. L.; Wu, Y. H.; Zhang, T. H.; Tu, L. H. Dihydrocaffeic Acid-Decorated Iron Oxide Nanomaterials Effectively Inhibit Human Calcitonin Aggregation. *ACS Omega* **2022**, *7* (35), pp. 31520–31528. https://doi.org/10.1021/ACSOMEGA.2C04206/ASSET/IMAGES/LARGE/AO2C04206_0007.JPEG.

Shivaji, K.; Sridharan, K.; Kirubakaran, D. D.; Velusamy, J.; Emadian, S. S.; Krishnamurthy, S.; Devadoss, A.; Nagarajan, S.; Das, S.; Pitchaimuthu, S. Biofunctionalized CdS Quantum Dots: A Case Study on Nanomaterial Toxicity in the Photocatalytic Wastewater Treatment Process. *ACS Omega* **2023**, *8* (22), pp. 19413–19424. https://doi.org/10.1021/acsomega.3c00496.

Shreve, M. J.; Brennan, R. A. Trace Organic Contaminant Removal in Six Full-Scale Integrated Fixed-Film Activated Sludge (IFAS) Systems Treating Municipal Wastewater. *Water Resources* **2019**, *151*, pp. 318–331. https://doi.org/10.1016/J.WATRES.2018.12.042.

Shukla, A. K.; Alam, J.; Mishra, U.; Alhoshan, M. A Sustainable Approach for the Removal of Pharmaceutical Contaminants from Effluent Using Polyamide Thin-Film Composite Membranes Integrated with Zn-Based Metal Organic Frameworks. *Environmental Science and Pollution Research* **2023**, *30* (51), pp. 110104–110118. https://doi.org/10.1007/S11356-023-30056-Z/METRICS.

Siddique, N.; Din, M. I.; Khalid, R.; Hussain, Z. A Comprehensive Review on the Photocatalysis of Congo Red Dye for Wastewater Treatment. *Reviews in Chemical Engineering* **2024**, *40* (4), pp. 481–510. https://doi.org/10.1515/REVCE-2022-0076/XML.

Singh, P.; Yadav, S. K.; Kuddus, M. *Green Nanomaterials for Wastewater Treatment;* Springer, Singapore, **2020**; Vol. 126. https://doi.org/10.1007/978-981-15-3560-4_9.

Soltani, S.; Gacem, A.; Choudhary, N.; Yadav, V. K.; Alsaeedi, H.; Modi, S.; Patel, A.; Khan, S. H.; Cabral-Pinto, M. M. S.; Yadav, K. K.; Patel, A. Scallion Peel Mediated Synthesis of Zinc Oxide Nanoparticles and Their Applications as Nano Fertilizer and Photocatalyst for Removal of Organic Pollutants from Wastewater. *Water (Basel)* **2023**, *15* (9), p. 1672. https://doi.org/10.3390/W15091672.

Som, I.; Roy, M.; Saha, R. Advances in Nanomaterial-Based Water Treatment Approaches for Photocatalytic Degradation of Water Pollutants. *ChemCatChem* **2020**, *12* (13), pp. 3409–3433. https://doi.org/https://doi.org/10.1002/cctc.201902081.

Subtil, E. L.; Almeria Ragio, R.; Lemos, H. G.; Scaratti, G.; García, J.; Le-Clech, P. Direct Membrane Filtration (DMF) of Municipal Wastewater by Mixed Matrix Membranes (MMMs) Filled with Graphene Oxide (GO): Towards a Circular Sanitation Model. *Chemical Engineering Journal* **2022**, *441*, p. 136004. https://doi.org/10.1016/J.CEJ.2022.136004.

Svensson Grape, E.; Chacón-García, A. J.; Rojas, S.; Pérez, Y.; Jaworski, A.; Nero, M.; Åhlén, M.; Martínez-Ahumada, E.; Galetsa Feindt, A. E.; Pepillo, M.; Narongin-Fujikawa, M.; Ibarra, I. A.; Cheung, O.; Baresel, C.; Willhammar, T.; Horcajada, P.; Inge, A. K. Removal of Pharmaceutical Pollutants from Effluent by a Plant-Based Metal–Organic Framework. *Nature Water* **2023**, *1* (5), pp. 433–442. https://doi.org/10.1038/s44221-023-00070-z.

Teodosiu, C.; Barjoveanu, G.; Sluser, B. R.; Popa, S. A. E.; Trofin, O. Environmental Assessment of Municipal Wastewater Discharges: A Comparative Study of Evaluation Methods. *The International Journal of Life Cycle Assessment* **2016**, *21* (3), pp. 395–411. https://doi.org/10.1007/S11367-016-1029-5.

Thakur, T. K.; Barya, M. P.; Dutta, J.; Mukherjee, P.; Thakur, A.; Swamy, S. L.; Anderson, J. T. Integrated Phytobial Remediation of Dissolved Pollutants from Domestic Wastewater through Constructed Wetlands: An Interactive Macrophyte-Microbe-Based Green and Low-Cost Decontamination Technology with Prospective Resource Recovery. *Water (Basel)* **2023**, *15* (22), p. 3877. https://doi.org/10.3390/W15223877.

Ulucan-Altuntas, K.; Debik, E. Dechlorination of Dichlorodiphenyltrichloroethane (DDT) by Fe/Pd Bimetallic Nanoparticles: Comparison with NZVI, Degradation Mechanism, and Pathways. *Frontiers of Environmental Science & Engineering* **2020**, *14* (1), pp. 1–13. https://doi.org/10.1007/S11783-019-1196-2/METRICS.

Villaseñor, M. J.; Ríos, Á. Nanomaterials for Water Cleaning and Desalination, Energy Production, Disinfection, Agriculture and Green Chemistry. *Environmental Chemistry Letters* **2017**, *16* (1), pp. 11–34. https://doi.org/10.1007/S10311-017-0656-9.

Wu, J.; Zhu, G.; Yu, R. Fates and Impacts of Nanomaterial Contaminants in Biological Wastewater Treatment System: A Review. *Water, Air, & Soil Pollution* **2017**, *229* (1), pp. 1–21. https://doi.org/10.1007/S11270-017-3656-2.

Xiao, W. D.; Xiao, L. P.; Lv, Y. H.; Yin, W. Z.; Sánchez, J.; Zhai, S. R.; An, Q. Da; Sun, R. C. Lignin-Derived Carbon Coated Nanoscale Zero-Valent Iron as a Novel Bifunctional Material for Efficient Removal of Cr(VI) and Organic Pollutants. *Separation and Purification Technology* **2022**, *299*, p. 121689. https://doi.org/10.1016/J.SEPPUR.2022.121689.

Yang, J.; Hou, B.; Wang, J.; Tian, B.; Bi, J.; Wang, N.; Li, X.; Huang, X. Nanomaterials for the Removal of Heavy Metals from Wastewater. *Nanomaterials* **2019**, *9* (3), p. 424. https://doi.org/10.3390/NANO9030424.

Yoon, S.; Cho, K. H.; Kim, M.; Park, S. J.; Lee, C. G.; Choi, N. C. Selenium Removal from Aqueous Solution Using a Low-Cost Functional Ceramic Membrane Derived from Waste Cast Iron. *Water (Basel)* **2023**, *15* (2), p. 312. https://doi.org/10.3390/W15020312.

Zamel, D.; Khan, A. U.; Waris, A.; Ebrahim, A.; Abd El-Sattar, N. E. A. Nanomaterials Advancements for Enhanced Contaminant Removal in Wastewater Treatment: Nanoparticles, Nanofibers, and Metal-Organic Frameworks (MOFs). *Results in Chemistry* **2023**, *6*, p. 101092. https://doi.org/10.1016/J.RECHEM.2023.101092.

Zazouli, M. A.; Eslamifar, M.; Javan, F. Water Disinfection Using Silver and Zinc Oxide Nanoparticles. *Journal of Nano Research* **2021**, *69*, pp. 105–121. https://doi.org/10.4028/WWW.SCIENTIFIC.NET/JNANOR.69.105.

Zeng, J.; Xia, Y. Not Just a Pretty Flower. *National Nanotechnology* **2012**, *7* (7), pp. 415–416. https://doi.org/10.1038/nnano.2012.105.

Zeng, M.; Chen, M.; Huang, D.; Lei, S.; Zhang, X.; Wang, L.; Cheng, Z. Engineered Two-Dimensional Nanomaterials: An Emerging Paradigm for Water Purification and Monitoring. *Materials Horizons* **2021**, *8* (3), pp. 758–802. https://doi.org/10.1039/D0MH01358G.

Zhang, Z.; Wu, J.; Yang, L.; Zhu, Y. Monitoring Effluent Water Quality in Wastewater Treatment Process Based on Multi-Model Online Soft Sensing Method. *8th International Conference on Image, Vision and Computing, ICIVC* **2023**, pp. 172–177. https://doi.org/10.1109/ICIVC58118.2023.10270767.

Zieliński, M.; Kazimierowicz, J.; Dębowski, M. Advantages and Limitations of Anaerobic Wastewater Treatment—Technological Basics, Development Directions, and Technological Innovations. *Energies* **2022**, *16* (1), p. 83. https://doi.org/10.3390/EN16010083.

Zimmer, A.; Winkler, I. T.; De Albuquerque, C. Governing Wastewater, Curbing Pollution, and Improving Water Quality for the Realization of Human Rights. *Waterlines* **2014**, *33* (4), 337–356. https://doi.org/10.3362/2046-1887.2014.034.

8 Challenges and Sustainable Solutions for the Detection and Bioremediation of Microplastic Pollution

Challenges and Scope of Their Remediation From the Environment

Rajni Sharma, Love Singla, and Gulab Singh

8.1 INTRODUCTION

Microplastics have infiltrated nearly every corner of the globe, from the deepest ocean trenches to remote mountain peaks. This widespread presence is primarily due to the durability and persistence of plastic materials, which can take hundreds of years to degrade fully. As more oversized plastic items break down into smaller particles through physical, chemical, and biological processes, they become increasingly difficult to detect and remove from the environment. Aquatic environments are particularly vulnerable to microplastic pollution. Marine organisms, ranging from tiny plankton to large mammals, can ingest these particles, leading to physical harm, toxic exposure, and disruption of biological functions. Microplastics can also be vectors for harmful chemicals and pathogens, further exacerbating their impact on marine life and food safety (Andrady and Anthony, 2011). The ingestion of microplastics by fish and other seafood raises concerns about human exposure through the food chain. Terrestrial ecosystems are not immune to the effects of microplastics. Soil contamination with these particles can affect soil structure, nutrient cycling, and the health of soil organisms. Moreover, microplastics can enter agricultural systems via contaminated water and fertilizers, potentially impacting crop health and food security.

Detecting and quantifying environmental microplastics is a critical first step toward addressing this pollution problem. However, several challenges hinder

DOI: 10.1201/9781003407317-8

effective detection. The small size of microplastics requires advanced analytical techniques, such as spectroscopy and microscopy, which are often time-consuming and expensive. Additionally, the diverse nature of microplastics, including variations in shape, size, color, and chemical composition, complicates their identification and analysis. Standardized sampling, extraction, and characterization methods are needed to improve the reliability and comparability of data across studies (Li et al., 2018). Using microorganisms to degrade or transform pollutants, bioremediation offers a promising and sustainable solution for microplastic pollution. Certain microbial strains have shown the ability to degrade plastic polymers, potentially breaking them down into less harmful substances. However, the efficiency of microbial degradation is limited by the recalcitrant nature of many plastics, which are designed to resist degradation. Enhancing the capabilities of microorganisms through genetic engineering and synthetic biology could improve the effectiveness of bioremediation processes. Despite its potential, bioremediation faces several challenges. The release of toxic byproducts during microbial degradation poses environmental and health risks that must be carefully managed. Additionally, the complex and varied nature of microplastic pollution requires a combination of bioremediation and other remediation strategies, such as phytoremediation and nanotechnology, to achieve comprehensive cleanup (Yan et al., 2022). Public awareness and regulatory frameworks are essential to successfully responding to microplastic pollution. Reducing plastic production and consumption, improving waste management systems, and promoting sustainable practices can help prevent further contamination. Collaborative efforts among scientists, policymakers, industries, and communities are crucial for developing and implementing effective detection and bioremediation strategies (Kuzma et al., 2023). Addressing microplastic pollution requires a multifaceted approach that includes advanced detection techniques, innovative bioremediation methods, and robust public and regulatory support. By tackling the challenges and leveraging sustainable solutions, we can mitigate the impact of microplastics on the environment and human health (Tedeschi et al., 2023).

8.2 CHALLENGES IN DETECTION OF MICROPLASTICS

Microplastic pollution presents a multifaceted challenge due to the diverse characteristics of these tiny particles. Detecting and quantifying microplastics in various environmental matrices is crucial for understanding their distribution, sources, and impacts. However, several factors complicate the detection process, including the size and diversity of microplastics, the complexity of analytical techniques required, and their widespread distribution across different environmental compartments (Prapanchan et al., 2023).

8.2.1 SIZE AND DIVERSITY OF MICROPLASTICS

Microplastics are plastic particles smaller than 5 mm, encompassing a wide range of shapes, sizes, colors, and polymer types. This heterogeneity poses significant challenges for detection and analysis. Microplastics can be fragments, fibers, beads, or films, each with different physical properties that affect their environmental behavior and interaction with detection methods. The small size of microplastics makes them difficult to isolate and identify. Particles less than 1 mm, often called micro-sized or sub-microplastics, are particularly challenging to detect due to their minuscule dimensions. The smaller the particles, the more sophisticated and sensitive the analytical techniques must be. Additionally, nano-plastics, particles smaller than 100 nano-meters, add another layer of complexity. These ultrafine particles require advanced detection methods, often at the limits of current technology. Microplastics also vary widely in their chemical composition, including various polymers such as polyethylene (PE), polypropylene (PP), polystyrene, and polyethylene terephthalate (PET), among others. Each polymer type may require different analytical approaches for accurate identification and quantification. Furthermore, microplastics can contain additives, colorants, and adsorbed pollutants, complicating their chemical analysis (Phipps et al., 2024).

8.2.2 ANALYTICAL TECHNIQUES FOR DETECTION

Detecting microplastics requires a combination of sampling, extraction, and analytical techniques, each with its own challenges. Sampling methods must be designed to capture a representative portion of microplastics from various environmental matrices, including water, sediment, soil, and air. Standard sampling techniques include using nets for aquatic environments and sediment corers or soil samplers for terrestrial matrices. However, ensuring that samples are free from contamination and represent the accurate concentration of microplastics in the environment is challenging.

Once samples are collected, the next critical step is to extract microplastics from the environmental matrix. This often involves separation techniques such as density separation, where microplastics are separated from heavier materials using liquids of different densities. However, this method may not effectively isolate all types of microplastics, particularly those with densities similar to the surrounding matrix.

Analytical techniques for identifying and quantifying microplastics include spectroscopy, microscopy, and thermal analysis. Fourier-transform infrared spectroscopy (FTIR) and Raman spectroscopy are commonly used to identify the chemical composition of microplastics. These techniques involve directing a beam of infrared or laser light at the sample and analyzing the resulting spectra to determine the polymer type. While highly effective, these methods can be time-consuming and require specialized equipment and expertise. Microscopy techniques, including optical microscopy, scanning electron microscopy (SEM), and transmission electron microscopy (TEM), are used to visualize microplastics and assess their size, shape, and surface characteristics. Optical microscopy suits larger microplastics, while SEM and TEM are necessary for examining smaller particles and nano-plastics. However, these techniques are labor-intensive and require meticulous sample preparation.

Thermal analysis methods, such as pyrolysis-gas chromatography-mass spectrometry (Py-GC-MS), involve heating the sample to break down the polymers into smaller molecules, which are then analyzed to identify the original plastic material. This technique is effective for determining the polymer composition of microplastics but requires sophisticated instrumentation and can be destructive to the samples. Combining these analytical techniques can provide a comprehensive understanding of microplastic pollution but also requires significant resources and expertise. Developing standardized sampling, extraction, and analysis methods is essential for improving data comparability and reliability across studies (Prapanchan et al., 2023).

8.2.3 DISTRIBUTION ACROSS ENVIRONMENTAL MATRICES

Microplastics are found in a wide range of environmental matrices, each presenting unique challenges for detection. Microplastics exist in surface waters, the water column, and sediments in aquatic environments. Detecting microplastics in surface waters often involves using nets with fine mesh sizes to capture floating particles. However, micro and nano-plastics may pass through the mesh, leading to underestimating their concentrations. Additionally, microplastics can be distributed throughout the water column, requiring specialized sampling devices that can collect samples at various depths.

Sediments pose a different challenge due to their complex and heterogeneous nature. Microplastics can be buried within the sediment matrix, making them difficult to extract and analyze. Density separation and sieving are commonly used techniques, but they may not be effective for all microplastics. Moreover, organic matter and other particulates can interfere with the extraction process and subsequent analysis. Soil contamination with microplastics is an emerging concern, particularly in agricultural areas where plastics are used extensively. Microplastics can enter the soil through contaminated water, fertilizers, and agricultural films. Detecting microplastics in soil requires sampling techniques that can capture particles from different soil layers. Extraction methods, such as density separation and filtration, need to be adapted to the specific properties of the soil matrix. The presence of organic matter and minerals can complicate the extraction and identification of microplastics.

Airborne microplastics represent another challenging matrix for detection. Wind and atmospheric currents transport these particles over long distances, eventually depositing them in remote areas. Sampling airborne microplastics typically involves the use of air filters or deposition samplers. However, distinguishing microplastics from other airborne particulates, such as dust and pollen, requires advanced analytical techniques. Microscopy and spectroscopy are often used to identify and quantify airborne microplastics, but these particles' small size and low concentrations make detection difficult.

The widespread distribution of microplastics across these diverse environmental matrices underscores the need for comprehensive and integrated detection strategies. Combining data from different matrices can provide a more complete picture of microplastic pollution and its sources. Collaborative efforts among researchers, policymakers, and industries are essential for developing effective detection methods and addressing the global challenge of microplastic pollution.

Detecting microplastics is a complex and multifaceted challenge that requires advanced analytical techniques, standardized methods, and collaborative efforts. Addressing the size and diversity of microplastics, improving sampling and extraction methods, and understanding their distribution across different environmental matrices are critical steps toward mitigating the impacts of microplastic pollution. By overcoming these challenges, we can develop effective strategies for monitoring and managing microplastics in the environment, ultimately protecting ecosystems and human health (Brewer et al., 2011).

8.3 BIOREMEDIATION OF MICROPLASTICS

Using microorganisms to degrade environmental pollutants, bioremediation is a promising strategy for addressing microplastic pollution. Microbial degradation of plastics harnesses the natural metabolic processes of bacteria, fungi, and other microorganisms to break down plastic polymers into more straightforward, less harmful substances. Despite its potential, bioremediation of microplastics faces several challenges, including microbial efficiency and the environmental and health risks of degradation byproducts.

8.3.1 Microbial Degradation of Plastics

Microbial degradation of plastics involves various microorganisms that can utilize plastic polymers as a carbon source. This process typically occurs in two main stages: biodeterioration and bioassimilation. During biodeterioration, microorganisms break down the plastic's surface, causing physical and chemical changes. Enzymes produced by these microorganisms, such as hydrolases, oxidases, and dehydrogenases, catalyze the breakdown of polymer chains into smaller molecules. The microorganisms take up these smaller molecules in the assimilation stage, where they are further metabolized into carbon dioxide, water, and biomass.

Bacteria and fungi are the primary microorganisms involved in plastic degradation. Several bacterial species, including those from the genera *Pseudomonas*, *Bacillus*, and *Rhodococcus*, have demonstrated the ability to degrade various plastic types, such as PE, PP, and polystyrene. Fungi, particularly from the genera *Aspergillus* and *Penicillium*, have also shown potential in breaking down plastic polymers. These microorganisms produce extracellular enzymes that can attack the chemical bonds in plastic materials, leading to their degradation.

One of the critical factors influencing microbial degradation is the structure and composition of the plastic. Biodegradable plastics, such as polylactic acid (PLA) and polyhydroxyalkanoates (PHAs), are designed to be more susceptible to microbial attack. However, conventional plastics, including PE and PP, are more resistant due to their high molecular weight and hydrophobic nature. The presence of additives, stabilizers, and plasticizers in these materials can further inhibit microbial activity (Kale et al., 2015).

8.3.2 Challenges in Microbial Efficiency

The efficiency of microbial degradation of plastics is limited by several factors.

a. Firstly, the intrinsic properties of plastics, such as their hydrophobicity and crystallinity, hinder the accessibility of microbial enzymes to the polymer chains. High molecular weight plastics with crystalline structures are particularly resistant to microbial attack, as their tightly packed molecules are less accessible to enzymatic action.

b. Secondly, the environmental conditions significantly affect microbial activity and plastic degradation rates. Factors such as temperature, pH, oxygen availability, and nutrient concentration influence the growth and metabolism of microorganisms. Optimal conditions for microbial degradation are often challenging to maintain in natural environments, leading to variable and usually slow degradation rates. In marine environments, for example, low temperatures and high salinity can inhibit microbial activity, reducing the effectiveness of bioremediation efforts.

c. Thirdly, the diversity and adaptation of microbial communities play a crucial role in the degradation process. While some microorganisms have evolved mechanisms to degrade plastics, others may lack the necessary enzymes or metabolic pathways. Enhancing microbial degradation efficiency requires identifying and engineering robust microbial strains capable of degrading a wide range of plastics under various environmental conditions. Genetic engineering and synthetic biology offer potential solutions by enabling the modification of microorganisms to enhance their plastic-degrading capabilities.

Furthermore, the formation of biofilms on plastic surfaces can either enhance or inhibit degradation. Biofilms are complex communities of microorganisms that adhere to surfaces and produce extracellular polymeric substances (EPS). While biofilms can facilitate the colonization and activity of plastic-degrading organisms, they can also create barriers that protect the underlying plastic from enzymatic attack. Understanding and manipulating biofilm dynamics is essential for improving microbial degradation efficiency.

8.3.3 Environmental and Health Risks of Byproducts

The microbial degradation of plastics can result in the formation of byproducts, some of which may pose environmental and health risks. During the breakdown of plastic polymers, intermediate degradation products, such as oligomers and monomers, can be released. These substances may be toxic to aquatic and terrestrial organisms, potentially disrupting ecosystems and food chains.

One of the major concerns is the release of micro and nano-plastics during the degradation process. As microorganisms degrade larger plastic particles, they can generate smaller fragments that persist in the environment. These micro nano-plastics can be ingested by various organisms, leading to physical and chemical impacts. Ingested microplastics can cause physical blockages, reduce feeding efficiency, and introduce harmful chemicals adsorbed on their surfaces into the organisms.

Moreover, the degradation of plastics can lead to the release of chemical additives used in plastic manufacturing. Additives such as plasticizers, stabilizers, flame retardants, and colorants can leach out during degradation, contaminating the environment. Many of these additives are known to be toxic, endocrine-disrupting, or carcinogenic, posing significant risks to wildlife and human health (Albazoni et al., 2024).

Another potential risk is the production of volatile organic compounds (VOCs) and greenhouse gases (GHGs) during microbial degradation. Specific microbial pathways can generate VOCs, contributing to air pollution and adverse health effects. Additionally, the incomplete degradation of plastics can release methane and carbon dioxide, potent GHGs that contribute to climate change.

To mitigate these risks, it is crucial to thoroughly assess the environmental and health impacts of microbial degradation byproducts. This includes understanding the fate and toxicity of intermediate degradation products, micro and nano-plastics, chemical additives, and gaseous emissions. Developing bioremediation strategies that minimize harmful byproducts while maximizing degradation efficiency is essential for safe and effective plastic pollution mitigation.

Understanding the factors influencing microbial degradation, developing robust microbial strains, and assessing the impacts of degradation byproducts are critical steps toward sustainable and effective bioremediation strategies. By leveraging advances in biotechnology and environmental science, we can harness the potential of microorganisms to mitigate the pervasive problem of microplastic pollution, protecting ecosystems and human health (Agarwal et al., 2024).

8.4 INNOVATIVE SOLUTIONS IN BIOREMEDIATION

Bioremediation of microplastics is a promising but complex endeavor that requires creative solutions to overcome the limitations of traditional methods. Recent advancements in genetic engineering, synthetic biology, and synergistic approaches involving phytoremediation and nanotechnology are opening new avenues for effectively addressing microplastic pollution.

8.4.1 GENETIC ENGINEERING AND SYNTHETIC BIOLOGY

Genetic engineering and synthetic biology offer powerful tools to enhance the capabilities of microorganisms in degrading microplastics. By manipulating the genetic makeup of bacteria and fungi, scientists can create strains with improved efficiency and specificity for breaking down plastic polymers. Genetic engineering involves inserting, deleting, or modifying genes within an organism to enhance its metabolic pathways for plastic degradation. For instance, genes encoding plastic-degrading enzymes, such as hydrolases and oxidases, can be introduced into microbial genomes to boost their ability to metabolize various plastic types. Researchers have successfully engineered strains of bacteria, like *Escherichia coli* and *Pseudomonas putida*, to produce enzymes that break down PET, a standard plastic used in bottles and textiles.

Synthetic biology goes a step further by designing and constructing entirely new biological systems or modifying existing ones in a highly controlled manner. This discipline allows for assembling novel metabolic pathways that can efficiently degrade plastics. For example, synthetic biologists can design microbial consortia where different microbial species with complementary metabolic capabilities work together to degrade complex plastic mixtures. These engineered consortia can be tailored to specific environmental conditions, enhancing their effectiveness in various settings.

Another promising approach in synthetic biology is using bioinformatics and computational modeling to predict and optimize enzyme functions. By analyzing the structure and activity of plastic-degrading enzymes, scientists can design more efficient variants that work under a broader range of conditions. This can lead to the development of robust microbial strains capable of degrading multiple types of plastics in diverse environments (Duncker et al., 2021).

8.4.2 Synergistic Approaches: Phytoremediation and Nanotechnology

Besides genetic engineering and synthetic biology, synergistic approaches combining bioremediation with phytoremediation and nanotechnology hold great potential for addressing microplastic pollution.

Phytoremediation involves the use of plants to absorb, accumulate, and degrade pollutants from the environment. Certain plants, known as hyperaccumulators, can take up microplastics from soil and water through their roots and leaves. Once absorbed, these microplastics can be broken down by plant-associated microorganisms or through the plants' own metabolic processes. Using genetically engineered plants that express plastic-degrading enzymes can further enhance this process. Phytoremediation offers a cost-effective and environmentally friendly solution for large-scale remediation of contaminated sites.

Nanotechnology provides innovative tools to improve the efficiency of microplastic degradation. Nanomaterials, such as nanoparticles and nanocomposites, can enhance the activity of plastic-degrading enzymes or facilitate the capture and removal of microplastics from the environment. For example, magnetic nanoparticles can be functionalized with plastic-degrading enzymes and dispersed in contaminated water. The nanoparticles can then be magnetically separated along with the degraded microplastic particles, providing a highly efficient cleanup method (Krzciuk et al., 2015).

Another application of nanotechnology is the development of nano-sensors for detecting microplastics. These sensors can monitor the presence and concentration of microplastics in real time, enabling more targeted and effective remediation efforts. Additionally, nanomaterials can be designed to adsorb and concentrate microplastics from large volumes of water, making subsequent degradation processes more efficient.

Combining bioremediation with phytoremediation and nanotechnology can create synergistic effects that enhance the overall efficiency of microplastic removal. For example, plants can be engineered to release nanomaterials that promote the activity of plastic-degrading enzymes in the rhizosphere, the soil region influenced by plant roots. This integrated approach can significantly increase the degradation rates of microplastics while minimizing environmental impact.

Innovative solutions in bioremediation, driven by advancements in genetic engineering, synthetic biology, and synergistic approaches involving phytoremediation and nanotechnology, hold great promise for addressing the global challenge of microplastic pollution. By enhancing the capabilities of microorganisms, leveraging the natural remediation potential of plants, and utilizing cutting-edge nanotechnologies, we can develop effective strategies to mitigate the impacts of microplastics on the environment and human health. These approaches represent a critical step forward in our efforts to create a cleaner, more sustainable future (Li et al., 2023).

8.5 ADVANCEMENTS IN ENVIRONMENTAL MONITORING

The detection and tracking of microplastic pollution have significantly advanced with new technologies and improved monitoring systems. These advancements enable more precise, efficient, and comprehensive assessments of microplastic distribution and impacts across various environmental matrices.

8.5.1 NEW TECHNOLOGIES FOR DETECTION

Several cutting-edge technologies have been developed to enhance the detection of microplastics in the environment. Among these, spectroscopy and imaging techniques stand out due to their accuracy and versatility.

a. FTIR spectroscopy and Raman spectroscopy are widely used to identify the chemical composition of microplastics. FTIR spectroscopy involves passing infrared light through a sample and analyzing the absorption spectrum to determine the material's molecular structure. On the other hand, Raman spectroscopy uses laser light to produce a spectrum based on the scattering of photons by the material's molecules. Both techniques can precisely identify different types of plastic polymers and even detect additives and contaminants associated with the microplastics (Li et al., 2018). Microplastics sensors represent another technological leap. These sensors are designed to provide real-time data on the presence and concentration of microplastics in water bodies. For example, optical sensors equipped with laser-induced fluorescence can detect microplastics by measuring the light emitted by plastic particles when exposed to a laser beam. Such sensors can be deployed in situ, offering continuous monitoring capabilities and reducing the need for labor-intensive sample collection and laboratory analysis (Käppler et al., 2015). Automated microscopy coupled with artificial intelligence (AI) also transforms microplastic detection. High-resolution microscopes can capture detailed images of environmental samples, and AI algorithms can be trained to identify and quantify microplastics within these images. This approach significantly speeds up the analysis process and increases accuracy by reducing human error (Mai et al., 2018).

b. Mass spectrometry techniques, such as Py-GC-MS, are also employed to analyze the thermal decomposition products of plastics, allowing for detailed characterization of microplastic composition. Although sophisticated and resource-intensive, these methods provide comprehensive data on the types and quantities of plastics in environmental samples (Dümichen et al., 2017).

8.5.2 Improved Tracking and Monitoring Systems

Advancements in tracking and monitoring systems are equally important in understanding and managing microplastic pollution. Integrating modern technologies into comprehensive monitoring frameworks enhances our ability to track microplastics' sources, pathways, and fate.

a. Geospatial technologies such as geographic information systems (GIS) and remote sensing are increasingly used to map the distribution of microplastics in various environments. GIS allows researchers to overlay spatial data on microplastic concentrations with other environmental and socio-economic data, facilitating the identification of pollution hotspots and potential sources. Remote sensing technologies, including satellite imagery and aerial drones, can cover large areas and provide valuable data on the spatial extent of microplastic pollution, especially in inaccessible or expansive regions (Guo et al., 2020).

b. Environmental DNA (eDNA) monitoring is another innovative approach. By analyzing DNA fragments in environmental samples, scientists can identify species that interact with or are affected by microplastics, offering indirect insights into the presence and impacts of microplastics in different ecosystems (Deiner et al., 2017).

c. Internet of Things (IoT) technology is also being leveraged to create intelligent monitoring networks. IoT devices with microplastic sensors can transmit real-time data to centralized databases, enabling continuous and automated monitoring. This networked approach allows timely responses to pollution events and supports long-term environmental monitoring efforts (Akyildiz et al., 2010).

d. Big data analytics and machine learning algorithms are crucial in processing and interpreting modern monitoring systems' vast amounts of data. These technologies can identify patterns and trends in microplastic pollution, predict future contamination scenarios, and inform policy-making and remediation strategies.

The advancements in environmental monitoring through new detection technologies and improved tracking systems are revolutionizing our ability to manage microplastic pollution. By providing more accurate, efficient, and comprehensive data, these innovations support adequate environmental protection and sustainable management practices (Shen et al., 2020).

8.6 PUBLIC AWARENESS AND REGULATORY FRAMEWORKS

Addressing microplastic pollution effectively requires a multifaceted approach involving public awareness, sustainable practices, plastic reduction strategies, and robust waste management systems. The synergy between public engagement and regulatory frameworks can drive significant progress toward mitigating this pervasive environmental issue.

8.6.1 ROLE OF PUBLIC AWARENESS

Public awareness is a cornerstone in the battle against microplastic pollution. Educating the public about the sources, impacts, and solutions to microplastic contamination fosters a collective responsibility and encourages behavioral changes that can significantly reduce plastic waste. Awareness campaigns can take various forms, including school programs, community workshops, media coverage, and social media initiatives.

Educational programs in schools play a vital role in shaping the younger generation's mindset. Integrating environmental science and sustainability into the curriculum can provide students with a solid understanding of plastic pollution's ecological and health impacts. Hands-on activities like beach cleanups and recycling projects can further reinforce these concepts and inspire long-term environmental stewardship (Hartley et al., 2018).

Community engagement is equally important. Workshops and public seminars can educate citizens about the importance of reducing plastic use, proper waste disposal, and the benefits of recycling. Local governments and non-governmental organizations (NGOs) can collaborate to organize these events, ensuring they reach a broad audience. Public campaigns using traditional and digital media can also raise awareness. Social media platforms, in particular, offer a powerful tool for spreading information quickly and engaging with a large audience. Influencers and environmental activists can leverage their platforms to highlight the issue of microplastics and advocate for sustainable practices.

Furthermore, citizen science initiatives can engage the public in data collection and environmental monitoring. By involving communities in scientific research, these programs raise awareness and empower individuals to contribute to the solution. Participants can collect samples from local environments, helping researchers track microplastic pollution and identify sources while gaining a deeper understanding of the issue (Hidalgo-Ruz and Thiel, 2015).

8.6.2 SUSTAINABLE PRACTICES AND PLASTIC REDUCTION

Adopting sustainable practices and reducing plastic consumption are critical steps toward mitigating microplastic pollution. Reducing the demand for single-use plastics and promoting alternatives can significantly decrease the amount of plastic waste entering the environment.

One of the most effective strategies is to encourage reusable products. Items like reusable bags, bottles, and containers can replace their single-use counterparts, reducing the overall plastic footprint. Public awareness campaigns can promote the benefits of reusable products, highlighting their environmental and economic advantages.

Another approach is to advocate for using biodegradable and compostable materials. These alternatives to conventional plastics can degrade more quickly and safely in the environment, minimizing the long-term impact of plastic waste. Supporting research and development in this area is crucial for creating viable, affordable alternatives that can be widely adopted (Singh and Sharma, 2016).

Governments and businesses also play a significant role in promoting sustainable practices. Implementing policies that ban or restrict single-use plastics can substantially reduce plastic waste. For example, several countries and cities have introduced bans on plastic bags, straws, and cutlery, encouraging consumers to switch to reusable or biodegradable options. Additionally, businesses can adopt sustainable practices by reducing plastic packaging, offering eco-friendly products, and implementing recycling programs.

Consumer behavior is another critical factor. Educating consumers about the environmental impact of their purchasing decisions can drive demand for sustainable products. Labels and certifications indicating environmentally friendly products can help consumers make informed choices. Encouraging consumers to support companies that prioritize sustainability can create market pressure for more businesses to adopt eco-friendly practices (Rujnić-Sokele et al., 2017).

8.6.3 WASTE MANAGEMENT SYSTEMS

Effective waste management systems are essential for controlling plastic pollution and preventing microplastics from entering the environment. Comprehensive waste management includes efficient collection, recycling, and disposal of plastic waste and the development of innovative technologies to improve these processes.

A crucial component of effective waste management is the implementation of robust recycling programs. Recycling not only reduces the amount of plastic waste but also conserves resources and energy. However, current recycling systems face several challenges, including contamination of recyclable materials, lack of infrastructure, and economic constraints. Improving public education on proper recycling practices can reduce contamination and increase recycling rates. Investing in advanced sorting and processing technologies can enhance the efficiency and effectiveness of recycling systems (Hopewell et al., 2009).

Extended producer responsibility (EPR) is a policy approach that holds manufacturers accountable for the lifecycle of their products, including post-consumer waste management. EPR programs can incentivize producers to design more sustainable products and packaging, reduce waste, and invest in recycling infrastructure. By shifting the responsibility of waste management from consumers to producers, EPR can drive significant improvements in plastic waste management (Agrawal et al., 2013).

Innovative technologies also offer solutions for improving waste management. Chemical recycling, for instance, involves breaking down plastic waste into its essential chemical components, which can then be used to produce new plastics. This process can handle a broader range of plastic types than traditional mechanical recycling and produce higher-quality recycled materials. Additionally, developing biodegradable plastics that can be safely broken down by natural processes can reduce the long-term impact of plastic waste.

Proper waste disposal is equally essential. Landfills and incineration are standard disposal methods but have significant environmental drawbacks. Landfills can leak pollutants into the soil and water, while incineration can release harmful emissions into the air. Implementing waste-to-energy technologies, which convert waste into usable energy, can provide a more sustainable alternative. These technologies can reduce the volume of trash, generate renewable energy, and minimize environmental impacts.

Government regulations and policies are crucial for ensuring adequate waste management. Regulations that mandate recycling, set waste reduction targets, and promote waste management infrastructure development can drive significant progress. International cooperation and agreements can also play a role in addressing plastic pollution on a global scale.

Public awareness and robust regulatory frameworks are essential for tackling microplastic pollution. By educating the public, promoting sustainable practices, and implementing effective waste management systems, we can reduce the impact of plastic pollution on the environment and human health. Collaborative efforts among governments, businesses, NGOs, and communities are crucial for creating a sustainable future free from the detrimental effects of microplastics (Ragaert et al., 2017).

8.7 COLLABORATIVE EFFORTS

Addressing microplastic pollution requires concerted efforts from multiple stakeholders, including scientists, policymakers, industries, and communities. Collaborative initiatives can leverage the strengths of each group to develop and implement effective solutions, fostering a holistic approach to this pervasive environmental problem.

8.7.1 INVOLVEMENT OF SCIENTISTS, POLICYMAKERS, AND INDUSTRIES

Scientists play a crucial role in understanding the extent and impact of microplastic pollution. Their research provides essential data on microplastics' sources, distribution, and effects on ecosystems and human health. By developing advanced detection methods and bioremediation technologies, scientists contribute to innovative solutions for mitigating microplastic contamination. Furthermore, interdisciplinary research involving chemists, biologists, and environmental scientists is critical to addressing the complex nature of microplastic pollution.

Policymakers are responsible for creating and enforcing regulations that limit plastic production and pollution. Effective policies can drive significant reductions in plastic waste through bans on single-use plastics, incentives for using biodegradable materials, and mandates for recycling. Policymakers must rely on scientific evidence to formulate regulations that are both effective and feasible. Collaborating with scientists ensures policies are based on the latest research and technological advancements, enhancing their impact (Galloway et al., 2016).

Industries have a significant role in reducing plastic pollution through sustainable practices and innovation. By adopting circular economy principles, companies can design products that minimize waste and are easier to recycle. Investments in research and development can lead to the creation of eco-friendly materials and packaging. Moreover, industries can implement EPR programs, taking accountability for the lifecycle of their products and encouraging recycling and proper waste management. Collaboration between industries and researchers can also lead to the development of scalable bioremediation technologies.

Public-private partnerships can further enhance these collaborative efforts. Joint initiatives involving governments, academic institutions, and businesses can pool resources and expertise to tackle microplastic pollution more effectively. These partnerships can fund research projects, support community education programs, and develop waste management and recycling infrastructure (Jambeck et al., 2015).

8.7.2 COMMUNITY ENGAGEMENT

Community engagement is vital for any initiative to reduce microplastic pollution. Public awareness campaigns can educate citizens about the environmental and health impacts of microplastics, encouraging more sustainable behavior. When individuals understand the importance of reducing plastic use, recycling properly, and supporting eco-friendly products, they are more likely to adopt practices that mitigate plastic pollution.

Local communities can also participate in citizen science projects, contributing valuable data to scientific research. Researchers can gain insights from a broader range of environments by involving the public in monitoring and data collection efforts while citizens develop a deeper connection to the issue. Community-led cleanup events, such as beach or river cleanups, reduce local plastic pollution, raise awareness, and foster a sense of responsibility (Hidalgo-Ruz and Thiel, 2015).

Educational institutions, from schools to universities, can play a pivotal role in community engagement. Integrating environmental education into the curriculum can raise awareness among young people, fostering a culture of sustainability from an early age. Workshops, seminars, and outreach programs can further educate the public about microplastic pollution and the actions they can take to address it.

NGOs and advocacy groups are essential in driving community engagement. These organizations can organize events, provide educational resources, and advocate for local, national, and international policy changes. NGOs can amplify their impact by collaborating with scientists, policymakers, and industries and ensuring that community voices are heard in decision-making processes.

The collaborative efforts of scientists, policymakers, industries, and communities are crucial for addressing microplastic pollution effectively. By working together, these stakeholders can develop innovative solutions, implement effective policies, and foster sustainable practices that mitigate the impact of microplastics on the environment and human health. We can make significant strides toward a cleaner and healthier planet through comprehensive and coordinated action (Kühn et al., 2015; Wabnitz et al., 2010).

8.8   FUTURE DIRECTIONS

The future of combating microplastic pollution hinges on advancing detection methods, improving bioremediation techniques, and implementing sustainable remediation strategies. Innovations in these areas will play a pivotal role in reducing the environmental and health impacts of microplastics.

8.8.1 Potential Developments in Detection and Bioremediation

The detection of microplastics is expected to benefit significantly from technological advancements. Emerging techniques such as hyperspectral imaging and advanced spectroscopy will enhance the accuracy and efficiency of identifying microplastics in various environmental matrices. Hyperspectral imaging can capture a wide range of wavelengths, allowing for precise differentiation between plastic particles and natural materials. Integrating AI and machine learning with these technologies can improve detection capabilities by automating the analysis process and reducing human error (Li et al., 2018).

Portable and real-time monitoring devices are another promising development. These devices can be deployed in various environments, providing continuous data on microplastic concentrations. For example, portable Raman and FTIR spectrometers are being miniaturized and adapted for field use, enabling on-site analysis of samples without the need for extensive laboratory equipment. This approach will facilitate more timely and targeted responses to microplastic pollution events (Shim et al., 2017).

Genetic engineering and synthetic biology are at the forefront of innovation in bioremediation. Researchers are exploring the potential of creating genetically modified microorganisms with enhanced plastic-degrading capabilities. These microorganisms can be tailored to produce specific enzymes that more efficiently break down a wide range of plastic polymers. Advances in CRISPR-Cas9 technology, a powerful tool for genetic editing, offer the possibility of designing microorganisms that can survive and thrive in diverse environmental conditions, further improving bioremediation effectiveness.

Synthetic biology also offers the potential to create entirely new metabolic pathways within microorganisms, enabling them to convert plastic waste into valuable byproducts such as biofuels or bioplastics. This approach helps in degrading plastics and promotes a circular economy by converting waste into valuable resources (Zhang et al., 2022).

8.8.2 Strategies for Sustainable Remediation

Implementing sustainable remediation requires a holistic approach combining technological advancements with policy measures and community engagement. One promising strategy is the integration of bioremediation with other remediation techniques, such as phytoremediation and nanotechnology. Plants engineered to express plastic-degrading enzymes can work with microorganisms to enhance the breakdown of microplastics in soil and water. Nanomaterials, such as enzyme-functionalized nanoparticles, can also accelerate the degradation process by targeting specific types of plastics and increasing the efficiency of microbial activity (Meng et al., 2020).

Promoting a circular economy is essential for sustainable remediation. This involves designing products and packaging with end-of-life disposal in mind, using materials that are easier to recycle or degrade. Policies that incentivize the use of biodegradable and recyclable materials and EPR schemes can drive industries to adopt more sustainable practices. EPR schemes make manufacturers responsible for the entire lifecycle of their products, encouraging the design of environmentally friendly products and ensuring proper waste management (Kirchherr et al., 2017).

Public education and community involvement are crucial components of sustainable remediation. Raising awareness about the impacts of microplastics and promoting sustainable consumer choices can significantly reduce plastic waste at the source. Community-based initiatives, such as local recycling programs and cleanup events, can foster a culture of environmental stewardship and support broader remediation efforts (Geyer et al., 2017).

Finally, international collaboration and policy harmonization are vital for addressing microplastic pollution on a global scale. Coordinated efforts between countries can lead to the development of standardized methods for monitoring and remediation, as well as the sharing of best practices and technologies. International agreements and regulations can help manage plastic production, reduce waste, and promote sustainable practices worldwide.

Future directions in detecting and bioremediating microplastics and sustainable remediation strategies offer promising solutions to this pressing environmental issue. By leveraging technological advancements, fostering collaboration, and promoting sustainable practices, we can mitigate the impacts of microplastics and move toward a cleaner, healthier planet (García-Marín and Rentería, 2024).

8.9   CONCLUSIONS

Addressing microplastic pollution requires a multifaceted approach that includes advanced detection techniques, innovative bioremediation methods, sustainable practices, and robust waste management systems. Continued research and collaboration among scientists, policymakers, industries, and communities are essential for developing and implementing practical solutions. By working together, we can mitigate the impacts of microplastics and move toward a cleaner, healthier planet. The ongoing commitment to research, innovation, and cooperation will be crucial in achieving long-term success in combating microplastic pollution.

REFERENCES

Agarwal, Tushar, Neeraj Atray, and Jai Gopal Sharma. "A critical examination of advanced approaches in green chemistry: microbial bioremediation strategies for sustainable mitigation of plastic pollution." *Future Journal of Pharmaceutical Sciences* 10 (2024): p. 78.

Agrawal, Vishal V., and Sezer Ülkü. "The role of modular upgradability as a green design strategy." *Manufacturing & Service Operations Management* 15 (2013): pp. 640–648.

Akyildiz, Ian F., and Mehmet Can Vuran. *Wireless sensor networks.* John Wiley & Sons Ltd, Sunrise Setting Ltd, Torquay, UK, 2010. https://content.e-bookshelf.de/media/reading/L-574273-88d4ba5edb.pdf.

Albazoni, Hamza Jasim, Mohammed Jawad Salih Al-Haidarey, and Afyaa Sabah Nasir. "A review of microplastic pollution: harmful effect on environment and animals, remediation strategies." *Journal of Ecological Engineering* 25 (2024): pp. 140–157.

Andrady, Anthony L. "Microplastics in the marine environment." *Marine Pollution Bulletin* 62 (2011): pp. 1596–1605.

Brewer, Shannon K., and Mark R. Ellersieck. "Evaluating two observational sampling techniques for determining the distribution and detection probability of age-0 smallmouth bass in clear, warmwater streams." *North American Journal of Fisheries Management* 31 (2011): pp. 894–904.

Deiner, Kristy, Holly M. Bik, Elvira Mächler, Mathew Seymour, Anaïs Lacoursière-Roussel, Florian Altermatt, Simon Creer et al. "Environmental DNA metabarcoding: transforming how we survey animal and plant communities." *Molecular Ecology* 26 (2017): pp. 5872–5895.

Dümichen, Erik, Paul Eisentraut, Claus Gerhard Bannick, Anne-Kathrin Barthel, Rainer Senz, and Ulrike Braun. "Fast identification of microplastics in complex environmental samples by a thermal degradation method." *Chemosphere* 174 (2017): pp. 572–584.

Duncker, Katherine E., Zachary Holmes, and Lingchong You. "Engineered microbial consortia: strategies and applications." *Microbial Cell Factories* 20 (2021): p. 211.

Galloway, Tamara S., and Ceri N. Lewis. "Marine microplastics spell big problems for future generations." *Proceedings of the National Academy of Sciences* 113 (2016): pp. 2331–2333.

García-Marín, Luis M., and Miguel E. Rentería. "Fighting plastic pollution with a circular economy roadmap and strategy: addressed to the United Nations Environment Programme." *Journal of Science Policy & Governance* 24(1) (2024): pp. 1–11.

Geyer, Roland, Jenna R. Jambeck, and Kara Lavender Law. "Production, use, and fate of all plastics ever made." *Science Advances* 3 (2017): e1700782.

Guo, Jing-Jie, Xian-Pei Huang, Lei Xiang, Yi-Ze Wang, Yan-Wen Li, Hui Li, Quan-Ying Cai, Ce-Hui Mo, and Ming-Hung Wong. "Source, migration and toxicology of microplastics in soil." *Environment International* 137 (2020): p. 105263.

Hartley, Bonny L., Sabine Pahl, Joana Veiga, Thomais Vlachogianni, Lia Vasconcelos, Thomas Maes, and Tom Doyle "Exploring public views on marine litter in Europe: perceived causes, consequences and pathways to change." *Marine Pollution Bulletin* 133 (2018): pp. 945–955.

Hidalgo-Ruz, Valeria, and Martin Thiel. "The contribution of citizen scientists to the monitoring of marine litter." *Marine Anthropogenic Litter* 16 (2015): pp. 429–447.

Hopewell, Jefferson, Robert Dvorak, and Edward Kosior. "Plastics recycling: challenges and opportunities." *Philosophical Transactions of the Royal Society B: Biological Sciences* 364 (2009): pp. 2115–2126.

Jambeck, Jenna R., Roland Geyer, Chris Wilcox, Theodore R. Siegler, Miriam Perryman, Anthony Andrady, Ramani Narayan, and Kara Lavender Law. "Plastic waste inputs from land into the ocean." *Science* 347 (2015): pp. 768–771.

Kale, S. K., Amit G. Deshmukh, Mahendra Shankarrao Dudhare, and Vikram Patil. "Microbial degradation of plastic - a review." *Journal of Biochemical Technology* 6 (2015): pp. 952–961.

Käppler, Andrea, Frank Windrich, Martin G. J. Löder, Mikhail Malanin, Dieter Fischer, Matthias Labrenz, Klaus-Jochen Eichhorn, and Brigitte Voit. "Identification of microplastics by FTIR and Raman microscopy: a novel silicon filter substrate opens the important spectral range below 1300 cm− 1 for FTIR transmission measurements." *Analytical and Bioanalytical Chemistry* 407 (2015): pp. 6791–6801

Kirchherr, Julian, Denise Reike, and Marko Hekkert. "Conceptualizing the circular economy: an analysis of 114 definitions." *Resources, Conservation and Recycling* 127 (2017): pp. 221–232.

Krzciuk, Karina, and Agnieszka Gałuszka. "Prospecting for hyperaccumulators of trace elements: a review." *Critical Reviews in Biotechnology* 35 (2015): pp. 522–532.

Kühn, Susanne, Elisa L. Bravo Rebolledo, and Jan A. Van Franeker. "Deleterious effects of litter on marine life." *Marine Anthropogenic Litter* (2015): pp. 75–116.

Kuzma, Jennifer, Khara D. Grieger, Ilaria Cimadori, Christopher L. Cummings, Nick Loschin, and Wei Wei. "Parameters, practices, and preferences for regulatory review of emerging biotechnology products in food and agriculture." *Frontiers in Bioengineering and Biotechnology* 11 (2023).

Li, Jingyi, Huihui Liu, and J. Paul Chen. "Microplastics in freshwater systems: a review on occurrence, environmental effects, and methods for microplastics detection." *Water Research* 137 (2018): pp. 362–374.

Li, Xiaona, Xiaowei Wang, Chunting Ren, Kumuduni Niroshika Palansooriya, Zhenyu Wang, and Scott X. Chang. "Microplastic pollution: phytotoxicity, environmental risks, and phytoremediation strategies." *Critical Reviews in Environmental Science and Technology* 54 (2023): pp. 486–507.

Mai, Lei, Lian-Jun Bao, Lei Shi, Charles S. Wong, and Eddy Y. Zeng. "A review of methods for measuring microplastics in aquatic environments." *Environmental Science and Pollution Research* 25 (2018): pp. 11319–11332.

Meng, Yuchuan, Frank J. Kelly, and Stephanie L. Wright. "Advances and challenges of microplastic pollution in freshwater ecosystems: a UK perspective." *Environmental Pollution* 256 (2020): p. 113445.

Phipps, Jesse, Bryant Passage, Kaan Sel, Jonathan Martinez, Milad Saadat, Teddy Koker, Natalie Damaso, Shakti Davis, Jeffrey Palmer, Kajal T. Claypool, Christopher Kiley, Roderic I Pettigrew, and Roozbeh Jafari. "Early adverse physiological event detection using commercial wearables: challenges and opportunities." *NPJ Digital Medicine* 7 (2024)

Prapanchan, Venkatraman Nagarani, Er. Vishal Kumar, T. Subramani, Udayakumar Sathya, and Peiyue Li. "A global perspective on microplastic occurrence in sediments and water with a special focus on sources, analytical techniques, health risks, and remediation technologies." *Water* 15 (2023): pp. 1987.

Ragaert, K., Delva, L., and Van Geem, K.. "Mechanical and chemical recycling of solid plastic waste." *Waste Management,* 69 (2017), pp. 24–58

Rujnić-Sokele, Maja, and Ana Pilipović. "Challenges and opportunities of biodegradable plastics: a mini review." *Waste Management & Research* 35 (2017): pp. 132–140.

Shen, Maocai, Shujing Ye, Guangming Zeng, Yaxin Zhang, Lang Xing, Wangwang Tang, Xiaofeng Wen, and Shaoheng Liu. "Can microplastics pose a threat to ocean carbon sequestration." *Marine Pollution Bulletin* 150 (2020): pp. 110712.

Shim, Won Joon, Sang Hee Hong, and Soeun Eo Eo. "Identification methods in microplastic analysis: a review." *Analytical Methods* 9(9) (2017): pp. 1384–1391.

Singh, P., and V. P. Sharma. "Integrated plastic waste management: environmental and improved health approaches." *Procedia Environmental Sciences* 35 (2016): pp. 692–700.

Tedeschi, Luis O. and Egleu D. M. Mendes. "13 precision livestock farming tools for climate-smart feedyard operations." *Journal of Animal Science* 101 (2023): pp. 326–327.

Wabnitz, Colette, and Wallace J. Nichols. "Plastic pollution: an ocean emergency." *Marine Turtle Newsletter* 129 (2010): pp. 1.

Yan, Muting, Xiaofeng Chen, Wei Chu, Weixin Li, Minqian Li, Zeming Cai, and Han Gong. "Microplastic pollution and enrichment of distinct microbiota in sediment of mangrove in Zhujiang River estuary, China." *Journal of Oceanology and Limnology* 41 (2022): pp. 215–228.

Zhang, Shengwei, Yanxia Li, Xingcai Chen, Xiaoman Jiang, Jing Li, Xiaoqi Yin, Liu Yang, and Xuelian Zhang. "Occurrence and distribution of microplastics (MPs) in organic fertilizers in China." *Science of the Total Environment* (2022): pp. 157061.

9 Challenges in Bioremediation
Overcoming Environmental and Technological Barriers

Harsha Singh, Suresh Kumar, and Atul Arya

9.1 INTRODUCTION

Bioremediation is a technique that uses microorganisms, plants, or enzymes to break down and remove environmental pollutants, thereby restoring contaminated sites. It holds immense potential for addressing the growing problem of environmental pollution in a sustainable manner. The processes in biological remediation transform hazardous substances, namely, polycyclic aromatic hydrocarbons (PAHs), heavy metals, and other hazardous compounds, into non-toxic or less hazardous compounds. A vast variety of microorganisms are used in the bioremediation process to detoxify, reduce, degrade, mineralize, or convert highly volatile or toxic contaminants into non-toxic or least toxic forms. Agrochemicals, dyes, hydrocarbons, heavy metals, plastics, radioactive waste, and other organic or inorganic pollutants are examples of pollutants whose properties greatly influence the efficacy of the restoration process. The amount and toxicity of chemicals present, as well as the site's characteristics and native microorganism population, all influence how well bioremediation technologies work at a given site (Azubuike et al., 2016). Sometimes, nutrition is added to treatment technologies in order to enhance the activity of native microorganisms. Microorganisms often flourish in ideal environmental conditions, and hazardous materials are effectively broken down as the number of microbial cells increases at the site. If the site lacks the essential biological activity, it is also feasible to import microorganisms from other areas that are ideal for decomposing a specific pollutant. Bioremediation is an environment-friendly approach to control pollution by utilizing biological agents, such as microbes and plants, to detoxify and restore contaminated sites. Despite its promise and potential, bioremediation faces numerous environmental and technological barriers that hinder its broader application and efficiency. Other bioremediation techniques include (i) phytoremediation, which uses plants to remove pollutants from soil and groundwater; (ii) phycoremediation, which utilizes macro- and microalgae to degrade pollutants; (iii) mycoremediation, which uses fungi like white-rot and others to reduce contaminants; and (Patel et al., 2022).

DOI: 10.1201/9781003407317-9

9.2 TECHNIQUES IN BIOREMEDIATION

Bioremediation employs various techniques to harness the natural abilities of organisms to detoxify and degrade environmental pollutants (Figure 9.1). These techniques can be broadly classified based on the type of organism used (microbes or plants), the environment in which they are applied (in situ/ex situ), and the specific strategies employed to enhance the remediation process.

9.2.1 In Situ Bioremediation Techniques

In situ bioremediation refers to the treatment of polluted groundwater or soil directly at the contamination site. This approach avoids the need for excavation or extraction, leveraging natural biological processes to degrade pollutants in place.

9.2.1.1 Intrinsic In Situ Bioremediation

Also known as natural attenuation or passive bioremediation, it refers to the natural degradation, transformation, or stabilization of contaminants in the environment through intrinsic biological processes without human intervention. Since there is no need for intervention, natural attenuation is an economical method, but still, the process needs to be monitored to ascertain its sustainability (Fragkou et al., 2021). The cycle depends on the strong and anaerobic microbial processes to break down harmful substances, especially those with strong personalities. The process is less expensive than other in situ technologies, as evidenced by the absence of external power (Hussain et al., 2022). This process relies on the inherent capabilities of native microbial communities and environmental conditions to reduce contaminant concentrations over time. The objective of this approach is the unsupported, passive

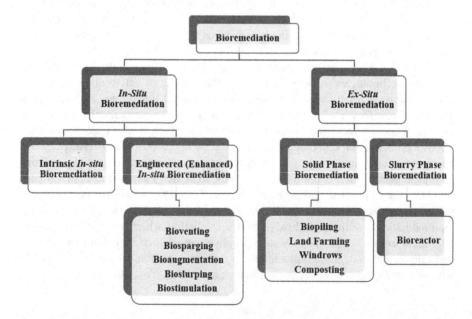

FIGURE 9.1   Different techniques used in bioremediation.

remediation of contaminated areas without any interference from humans. This method uses both anaerobic and aerobic microbial activity for the treatment of both biodegradable and non-biodegradable pollutants (Muttaleb and Ali, 2022). For the treatment of biodegradable and resistant pollutants, this method combines anaerobic and aerobic microbial activities (Nayak and Solanki, 2022). Intrinsic bioremediation involves various natural attenuation processes:

- **Biodegradation:** Microorganisms metabolize contaminants, transforming them into harmless end products like carbon dioxide, water, and biomass.
- **Dispersion:** Physical spreading of contaminants, reducing concentrations.
- **Dilution:** Mixing with uncontaminated water or soil, lowering contaminant levels.
- **Sorption:** Contaminants adhere to soil particles, reducing their mobility.
- **Volatilization:** Evaporation of volatile contaminants into the atmosphere.
- **Chemical Reactions:** Abiotic processes that chemically transform contaminants.

9.2.1.2   Engineered (Enhanced) In Situ Bioremediation

Engineered in situ bioremediation enhances natural biodegradation processes through deliberate human intervention, improving the efficiency as well as the effectiveness of the contaminant degradation process. This approach involves modifying the environmental conditions or introducing contaminant-specific microbial cultures to accelerate the breakdown of pollutants directly at the contaminated site.

- **Bioventing:** Bioventing technique uses microorganisms to decompose organic materials that adsorb on soil. This technique increases the microbial activity by supplying oxygen to the unsaturated zone. This facilitates the process of bioremediation, lowering carbon dioxide levels as a result (Ghosh et al., 2023). In order to improve the bioventing process, nutrients and moisture are also supplied during bioremediation. The only factor influencing gas exchange via the vent wells in passive bioventing technology is atmospheric air pressure. A blower is used in active bioventing to force air into the ground (Zouboulis and Moussas, 2019).
- **Biosparging:** Biosparging technique employs native microorganisms that break down organic materials in the saturation zone. Using this method, the saturation zone is injected with oxygen and nutrients (if required) to enhance the biological activity of native microorganisms (Muley et al., 2023). Biosparging works better for naturally remediating petroleum hydrocarbon compounds spilled in contaminated aquifers, such as ethylbenzene, xylene, diesel, benzene, toluene, and lamp fuel. The permeability of the soil and the biodegradability of contaminants are the two main factors that influence the effectiveness of biosparging. The use of biosparging is typically observed at locations that contain medium-weight petroleum products, such as diesel or jet fuel (Pande et al., 2020).
- **Bioaugmentation:** The term "polishing-up" or "finishing" refers to the process of bioaugmentation, which occurs when a fresh oil spill's initial high

concentration prevents it from quickly transforming into less hazardous components. In order to bioremediate contaminated water, bioaugmentation techniques are used for total petroleum hydrocarbon (TPH), crude oil, and related petroleum pollutants (Sayed et al., 2021). Researchers typically use *Pseudomonas aeruginosa* and *Bacillus subtilis* for bioaugmentation. In addition to the pollutant's type, concentration, bioavailability, and recalcitrance, the bioaugmentation process also depends on the microbial system and environment. It is capable of breaking down an extremely hazardous and resistant substance within an ecosystem, which may be impossible to fully mitigate by natural attenuation (Lopes et al., 2022).

- **Biostimulation:** The most common method of biostimulation involves incorporating specific nutrients into the polluted areas. The primary benefit of utilizing indigenous microorganisms in the biostimulation procedure is their excellent adaptability and subsurface distribution, which helps to avoid any issues regarding environmental incompatibility (Dehnavi and Ebrahimipour, 2022). Agarry and Ogunleye studied the effectiveness of three independent biostimulant variables—pig manure (organic nutrient), Tween 80 (nonionic surfactant), and Nitrogen, phosphorus, potassium (NPK) fertilizer (N:P:K ratio of 20:10:10) (inorganic nutrient). The study took 42 days to biodegrade TPH and hexavalent chromium (Cr (VI)) contaminants in order to remediate contaminated spent engine oil in the soil. TPH and Cr(VI) had the greatest biodegradation rates of 66.5% and 52.3%, respectively. At 4.22 g, 47.76 g, and 10.69 mg/L, respectively, for NPK fertilizer, pig dung, and Tween 80, the highest biodegradation rate of 66.5% and 52.3% for TPH and Cr(VI) was noted. The biostimulation of the native microbial community was the primary factor responsible for both the increase in biodegradation and the decrease in duration of biodegradation of the heavy metal co-contaminated soil and petroleum hydrocarbon, as compared to the control case's biodegradation rate of 28.8% and 10.4% for TPH and Cr(VI), respectively (Agarry and Ogunleye, 2012).

- **Bioslurping:** Bioslurping is an innovative in situ remediation technique that combines the best features of bioventing and vacuum-enhanced pumping in order to recover free product from soil and groundwater while activating aerobic bioremediation of hydrocarbon contaminants (Mapelli et al., 2017). A slurp tube of varying lengths is introduced into one or more wells in the bioslurping system. The slurp tube is attached to the vacuum pump that descends into a light non-aqueous phase of liquid layer to remove some groundwater and free product (Akhtar et al., 2023). The oil/water separator receives all the liquids (groundwater and product solution) removed from a slurp tube, whereas the liquid/vapor separator receives the vapors (Bodor et al., 2020).

9.2.2 Ex Situ Bioremediation Techniques

In ex situ bioremediation, materials are broken down by native microbial populations by placing excavated soil in a treatment area that is lined aboveground and allowing it to breathe after treatment. Usually, it is done for soil and sediments that have been removed from their original location and transferred to another location for treatment (Nag et al., 2024). Ex situ bioremediation strategies rely on variables such as the type of pollutant, its depth, treatment cost, degree of pollution, geology, and geographic location of the contaminated areas. Performance criteria also help in identifying ex situ bioremediation strategies.

9.2.2.1 Solid Phase Bioremediation

It involves the use of natural systems to degrade contaminants from sewage sludge and municipal solid wastes. It utilizes microbial activity to break down pollutants in soil or sediment without needing to transport the contaminated material. This method is cost-effective and environmentally friendly, suitable for treating various organic and inorganic contaminants.

- **Biopiling:** It involves stacking excavated soils in a treatment area with an aeration system. This method involves digging up polluted soil, stacking it in small piles, and then allowing the air movement through it to promote microbial activity that will break down the pollutants. A possible approach is to treat volatile contaminants with low molecular weight and clean up polluted environments using biopile (Sutherland and Ralph, 2019). The use of bulking agents like wood chips, sawdust, or straw has aided in the biopile's remediation process. Land farming, biosparging, and bioventing are examples of ex situ bioremediation procedures that can be used to restore the oxygen supply to polluted soil piled in biopiles (Arora et al., 2022). Biopile devices have been used for treating diesel-contaminated soil in the sub-Antarctic region. 93% of the TPH can be removed in less than 1 year using a biopile system (Yap et al., 2021).
- **Land Farming:** Land farming uses agricultural practices to promote natural pollutant biodegradation. Because it requires no specialized equipment and has cheap operational expenses, land farming is the most crucial and straightforward technique of bioremediation (Janssen and Stucki, 2020). Using this technology, the bioremediation of contaminated soil is a relatively simple process that requires minimal energy, little capital, and has very less ecological impact (Wang et al., 2021). In the land farming method, contaminated soil is routinely removed, dispersed, and tilled over the permanent soil layer support above ground to allow autochthonous bacteria to break down contaminants through aerobic biodegradation. In the Niger Delta, fertilizers have also been employed to improve the productivity of land cultivation on hydrocarbon-contaminated soil. PAHs are removed using this economical, environmentally beneficial technique. Bioremediation of PAH-contaminated soil has been demonstrated by data gathered from small-scale experiments conducted in the laboratory on integrated land farming and composting processes, showing very feasible results (Lukić et al., 2017).
- **Windrows:** In order to improve bioremediation by enhancing the decomposition activities by microbes of native and transitory hydrocarbonoclastic found in polluted soil, windrow bioremediation systems rely on periodically

rotating the piled polluted soil (Hussain et al., 2022). The pace of bioremediation can be effectively accelerated through acclimatization, mineralization, and biotransformation. Windrows are periodically turned to aerate the
soil, promoting microbial degradation of pollutants. The process enhances
oxygen flow and maintains optimal moisture and temperature levels, facilitating the breakdown of organic contaminants such as hydrocarbons, pesticides, and industrial chemicals. In comparison to biopile therapy, windrow
therapy showed an increased rate of hydrocarbon evacuation. Because of
the more erodible soil category, the hydrocarbon expulsion wind's increased
efficiency did not increase over time (Sharma et al., 2018).

- **Composting:** Composting is the process that produces compost, a humic
  supplement used as a soil fertilizer, by breaking down organic waste in the
  presence of thermophilic biological organisms (Narayan Chadar, 2018).
  In order to improve mixture permeability and aeration, which allows the
  exposure of more surface and redistributing the matter, compost is often
  arranged in long, tapered, and low piles. While this method is generally
  thought to be the least expensive, it is the least effective in terms of temperature control and aeration (Cristorean et al., 2016).

9.2.2.2 Slurry Phase Bioremediation

Slurry phase bioremediation is a reasonably quick process when compared to other
treatment techniques. Here, contaminated soil is mixed with water and additional
chemicals, and the mixture is then poured into a large vessel known as a bioreactor.
This allows the mixture to be sorted in a solid-liquid suspension, which brings in
native microorganisms.

- **Bioreactor:** The time required for bioremediation can be significantly reduced
  by using an effective bioreactor-based method that can control temperature,
  agitation, substrate concentration, pH, aeration, and inoculum concentration
  very precisely (Davoodi et al., 2020). Controllable and manipulable bioreactors allow biological reactions to occur. Designs of the bioreactor are flexible
  enough to maximize microbial decomposition while minimizing abiotic losses
  (Bala et al., 2022). Because a bioreactor can regulate, control, and alter the
  parameters in the process necessary for the biological processes involved in
  the biodegradation of pollutants, it is far more advantageous to employ one for
  the bioremediation process than to use another method (Azubuike et al., 2016).
  Contaminated media have been treated in slurry phase bioreactors. For soil
  and sediment samples contaminated with petroleum hydrocarbons, herbicides,
  insecticides, creosotes, tars, and chlorinated solvents, this bioreactor offers an
  environment-friendly approach for ex situ bioremediation technique. There
  are several varieties of bioreactors available, including continuous, batch,
  fed-batch, and multistage reactors. The cost of the bioreactor must be considered while choosing the type or mode to be utilized. Bioreactors make it easy
  to characterize the microorganisms and monitor changes in the population of
  microbes because they may be run in batches or for extended periods of time
  (Etuk et al., 2024).

9.2.3 MICROBIAL BIOREMEDIATION

Microorganisms can oxidize, reduce, methylate, demethylate, and complexate heavy metals to affect their biotransformation. The way that microbial activity impacts the accumulation and degradation of heavy metals depends on the physicochemical properties of metal complexes as well as the geochemical factors of contaminated areas (Dell'Anno et al., 2020). Bacteria are distinguished from other microbial forms by their toughness, availability of more surface area, rapid growth rate, and strong resistance to harmful heavy metals. Numerous mechanisms, such as redox reactions, complex forms of efflux pumps, extra- and intracellular sequestrations, and, in specific conditions, the utilization of the metal ions as electron acceptors in oxidative phosphorylation, have been developed by the bacterial strains to manage heavy metal degradation (Nanda et al., 2019). Mycoremediation is an affordable process that doesn't generate any hazardous waste. The fungus life cycle, the chemical behavior of the elements, and the presence or absence of the fungi after sequestration all affect the future accessibility to heavy metals and other contaminants. Fungal isolates that have been exposed to soil or wastewater contaminated with heavy metal frequently display resistance genes linked to heavy metal remediation (Jeyakumar et al., 2022). Microalgae require trace elements from heavy metals like cobalt, iron, boron, molybdenum, copper, manganese, and zinc for cellular metabolism and enzymatic processes. However, microalgae are toxic to heavy metals such as Cr, Pb, As, Cd, and Hg. Table 9.1 enlists the different microbes utilized for removing different types of pollutants present in contaminated soil or water.

TABLE 9.1
Microbes Used for Bioremediation

Type of Microorganism	Microbial Species	Remediated Pollutant	Reference
		Bacteria	
	Bacillus sp. and *Staphylococcus sp.*	Endosulfan	Liu et al. (2018)
	Pseudomonas aeruginosa, Alcaligenes odorans, Corynebacterium propinquum, and *Bacillus subtilis*	Phenol	Gaur et al. (2018)
	Microbacterium sp., Shigella sp. Micrococcus sp., and *Bacillus sp.*	Arsenic and Uranium	Bhakat et al. (2019)
	Fusarium sp., *C. propinquum, P. aeruginosa,* and *A. odorans*	Oils	Pande et al. (2020)
	Lysinibacillus sphaericus CBAM5	Lead, chromium cobalt, copper, and crude oil	Kharangate-Lad and D'Souza (2021)

(Continued)

TABLE 9.1 (*Continued*)
Microbes Used for Bioremediation

Type of Microorganism	Microbial Species	Remediated Pollutant	Reference
	Bacillus licheniformis	Dyes	Mousavi et al. (2021)
	Bacillus sp., Enterobacter sp. Escherichia sp., and *Shewanella sp.*	Chromium	Mousavi et al. (2021)
	Pseudomonas putida YNS1	Phenol, Cadmium, and Copper	Mehrotra et al. (2021)
	B. licheniformis JUG GS2 (MK106145) and *Bacillus sonorensis*	Naphthalene	Rabani et al. (2022)
	Cyclotella cryptica, Pseudochlorococcum typicum	Mercury	Shah and Jain (2020)
	P. cepacia, Serratia ficaria, Bacillus coagulans, and *Bacillus cereus*	Diesel oil	Miri et al. (2022)
Algae			
	Chlamydomonas reinhardtii and *S. almeriensis*	Arsenic	Saavedra et al. (2018)
	Cystoseira indicant	Cadmium and Nickel	Moreira et al. (2019)
	Spirogyra hyaline and *C reinhardtii*	Mercury	Shah and Jain (2020)
	Chlorella sp. and *Spirulina sp.*	Lead and nickel	Geetha et al. (2021)
	Chlorococcum humicola	Cobalt	Chugh et al. (2022)
Fungi			
	Phomopsis liquidambari	Phenanthrene	Fu et al. (2018)
	Coprinus comatus	4-Hydroxy-3,5-dichlorobiphenyl	Li et al. (2018)
	Saccharomyces cerevisiae	Arsenic	Duc et al. (2021)
	Trichoderma sp., Aspergillus sp., and *Penicillium sp.*	Cobalt and copper	Dusengemungu et al. (2020)
	Alternaria alternata, Aspergillus flavus, Aspergillus terreus, and *Trichoderma harzianum*	Polycyclic aromatic hydrocarbon	Prakash et al.

9.3   ENVIRONMENTAL BARRIERS

9.3.1   Nature of Contaminants

- **Complexity and Diversity of Pollutants:** The effectiveness of bioremediation is greatly influenced by the complexity and type of contaminants present at a site. Contaminants such as heavy metals, PAHs, polychlorinated biphenyls (PCBs), and various chlorinated compounds often exhibit recalcitrance, making them resistant to microbial degradation. For instance, heavy metals are not biodegradable and require different strategies, such as bioaccumulation or phytoremediation, to be removed or stabilized. Organic pollutants like PAHs and PCBs, due to their complex molecular structures, are difficult to break down and may resist in the environment for longer durations (Chandra et al., 2013).
- **Mixtures of Contaminants:** Environmental pollution rarely involves a single type of contaminant; more often it comprises complex mixtures. These mixtures can interact synergistically or antagonistically, affecting microbial degradation processes. Some contaminants may inhibit the activity of microorganisms or their enzymes, while others may serve as co-substrates, increasing the degradation of the primary pollutant (Lamichhane et al., 2016).

9.3.2   Site-Specific Conditions

- **Physical Characteristics of the Site:** The physical characteristics of a contaminated site, including soil texture, structure, porosity, and permeability, play a significant role in bioremediation. The movement of water, nutrients, and oxygen into the zone of biological activity is controlled by the texture of the soil. These elements are transported slowly by the fine soil particles, such as clay and silt. Sand and gravel-containing permeable soils are good for nutrient transport and can be remedied somewhat rapidly (Geng et al., 2022). Soil with high clay content, for example, may impede the movement of water and nutrients, limiting microbial activity. Conversely, sandy soils with high permeability may enhance the transport of pollutants but also pose challenges in maintaining adequate moisture and nutrient levels (Polyak et al., 2018). The majority of petroleum hydrocarbons are strongly adsorbed onto soil particles, especially clays, and the pace at which these hydrocarbons desorb from the soil is thought to be limited during the biodegradation process. Recently, there has been a suggestion about utilizing natural infiltration processes in surface foam spraying technologies to keep contaminated soils from being mixed or disturbed. However, it can be difficult to get reactive remedial materials to the bottom of a polluted zone using the mechanism of natural infiltration only (Jabbar et al., 2022).
- **Environmental Parameters:** Several environmental factors, such as temperature, pH, moisture content, and oxygen availability, critically affect microbial activity and consequently, bioremediation efficiency.

- **Temperature:** An increase in temperature also enhances microbial activity up to a certain point and declines after that. Extreme temperatures, either too high or too low, can inhibit microbial metabolism and enzyme functions necessary for pollutant degradation. At low temperatures, the volatility of hazardous low-molecular-weight hydrocarbons decreases, while the viscosity of petroleum compounds increases. This results in a delay in the biodegradation process (Svinterikos et al., 2019). Raising the temperature of hydrophobic pollutants will cause long-chain n-alkanes to become more soluble, less viscous, and increase their mass transfer from the solid to the liquid phase (Chen et al., 2020).
- **pH:** The pH in the environment influences the solubility and availability of nutrients and pollutants, as well as microbial community structure and function. Most bioremediation processes are optimal at near-neutral pH (between 6 and 9), where the microorganisms perform degrading functions (Bajagain et al., 2021). Natural environments have pH values between 5.0 and 9.0, which are ideal for the biodegradation of contaminants. The presence of carbonates and other minerals in most natural soils maintains this pH range (Bajagain et al., 2020).
- **Moisture Content:** The right amount of moisture influences microbial growth and activity. Water facilitates the transport of nutrients and pollutants to microbial cells. However, excessive moisture can lead to anaerobic conditions, unfavorable for aerobic microorganisms typically involved in bioremediation of organic pollutants. The removal rate is very slow when there is little or no water, which leads to a small number of organisms (Chaudhary et al., 2019). For plants, a moisture saturation range of 30%–80% is ideal for aerobic bioremediation to work well (Larik et al., 2016).
- **Oxygen Availability:** Oxygen is crucial for aerobic degradation processes. In oxygen-limited environments, anaerobic microorganisms may take over, but they generally degrade pollutants more slowly and incompletely compared to aerobic microbes. Oxygen can improve the metabolism of hydrocarbons (Kästner and Miltner, 2016). Balancing oxygen levels is particularly challenging in subsurface and water-saturated environments. Petroleum hydrocarbons can degrade up to 90% in an aerobic environment, but only up to 25% in an anaerobic environment (Naseri et al., 2014).

9.4 BIOAVAILABILITY OF POLLUTANTS

9.4.1 Soil and Sediment Adsorption

The pollutants' bioavailability to microorganisms for degradation is a significant barrier in bioremediation. Pollutants often adsorb onto soil and sediment particles, reducing their availability to microbial cells. Various factors such as organic matter content, soil texture, and the chemical nature of the pollutant influence adsorption and desorption dynamics. In order to improve bioavailability and the metabolism of contaminants, biosurfactants are used to make the substrate more soluble (Ma et al., 2015).

- **Soil Organic Matter:** High organic matter content can sequester hydrophobic pollutants, making them less bioavailable. Conversely, organic matter can also provide nutrients and binding sites that may enhance microbial activity.
- **Pollutant Solubility:** Hydrophobic pollutants, due to their low solubility in water, tend to partition into soil organic matter or sediment particles, reducing their accessibility to aqueous phase microorganisms.

9.4.2 NON-AQUEOUS PHASE LIQUIDS (NAPLs)

Pollutants present as non-aqueous phase liquids (NAPLs), such as petroleum hydrocarbons, present a significant challenge due to their limited solubility and high viscosity. NAPLs can form persistent contaminant sources, slowly releasing pollutants over time, thereby prolonging the contamination and complicating bioremediation efforts. The environment and human health are the main concerns associated with subsurface NAPL accumulation, and these hazards are becoming more important to society as a result of the ongoing climate change (Mineo, 2023).

9.5 ECOLOGICAL IMPACTS

Disruption of Local Ecosystems: Introducing foreign microorganisms or modifying environmental conditions to enhance bioremediation can disrupt local ecosystems. For example, bioaugmentation, the addition of specific microbial strains, may lead to competition with native microorganisms, potentially disrupting the existing microbial community balance. Similarly, biostimulation, which involves adding nutrients or electron acceptors, can alter nutrient dynamics and affect non-target organisms.

Potential for Pathogen Proliferation: Altering environmental conditions to favor bioremediation can sometimes inadvertently promote the growth of opportunistic pathogens. For instance, increasing nutrient levels or creating anaerobic conditions might encourage the proliferation of harmful microorganisms, posing health risks to humans and wildlife.

Ecosystem Recovery and Resilience: The long-term ecological impacts of bioremediation must be considered. While the primary goal is to remove contaminants, the process should also support the recovery and resilience of the ecosystem. This involves ensuring that bioremediation activities do not cause secondary pollution or long-lasting ecological disturbances.

9.6 TECHNOLOGICAL BARRIERS

9.6.1 BIOREMEDIATION TECHNIQUES

Several bioremediation techniques, including in situ and ex situ methods, have been developed. Each method has its limitations; in situ methods often face difficulties with uniform treatment and monitoring, whereas ex situ methods can be costly and logistically challenging. One of the main challenges in the bioaugmentation process is preparing the inoculum and getting it to the contamination site without diminishing its efficacy. Furthermore, because the microbial inoculum might not be able to adapt to the target site environment, the experimental capacity of microbial strains to digest a pollutant under laboratory conditions does not transfer to the natural environment. A number of additional variables also come into play, including pH, temperature, habitat, and microbial composition. Another drawback is that the introduction of pathogenic bacteria into the environment may be impacted by the presence of alternate carbon and sources of nutrients (Chettri et al., 2024). The bioaugmentation process is also constrained by nutrient shortages. Consequently, a number of studies have indicated that bioaugmentation and biostimulation could be combined to boost effectiveness. Furthermore, studies have demonstrated that bioaugmentation lowers microbial diversity by displacing other microbes and establishing the microbial inoculum as the dominant species in the hydrocarbon-contaminated soil. Compared to the biostimulation process, which involved the microbial communities cooperating to eradicate the contamination with nearly twice the efficacy of the bioaugmentation, this led to a decreased efficiency (Wu et al., 2019). A variety of pollutants can impact the microbial inoculum that targets a particular component in a single strain-based inoculum in terms of growth and biodegradation processes. In particular, a single contaminant may limit the bioaugmentation of another contaminant. Thus, it is desirable to have consortia of microbes that offer the power and metabolic variety required to quicken the process (Tyagi et al., 2011).

9.6.2 CHALLENGES IN MICROBIAL BIOREMEDIATION

Numerous factors, such as the chemical forms of the metals, redox potential, total metal ion concentration, competition between microbes, soil structure, moisture content, temperature, oxygen content, pH, nature of the soil, and the solubility of the heavy metal in water, influence the activity of microbes to biodegrade heavy metals. Microbes compete with one another for carbon, a limiting resource that they use as fuel (Ayilara and Babalola, 2023). The biodegradation rate would therefore be influenced by the innate capacity of the bacteria, which have a greater capacity to digest heavy metal pollutants. The type and population of microbes play a major role in the rate and effectiveness of a bioremediation process. For example, a particular strain of organism may be able to bioremediate a specific heavy metal, but in mass remediation, a complete microbial consortium would be required (Patel et al., 2022). Nitrogen (N) and phosphorus (P) are also needed by the bacteria that cause biodegradation; thus, it's necessary to balance the C:N:P ratio in an environment where these vital nutrients are scarce in order to enhance the process of biodegradation. The rate of microbial biodegradation is quite low with high moisture content. An efficient microbial biodegradation process needs an ideal moisture content, as bacteria may not be able to live at low moisture levels. In chilly areas where psychrophiles are the sole inhabitants, heavy metals are broken down slowly by microbes. This is due to the fact that, although most bioremediation processes are favored by high temperatures, the breakdown of each molecule also happens at different temperatures. This is because the microbial transport channels are frozen by sub-zero water, which reduces metabolic activity (Bala et al., 2022).

9.6.3   LIMITATIONS OF CURRENT TECHNOLOGIES

Current bioremediation technologies may not effectively address all types of contaminants or site conditions. For instance, biostimulation (enhancing microbial activity by adding nutrients) and bioaugmentation (introducing specific microbial strains) may not always result in the complete degradation of pollutants. The primary drawback of bioslurping is the excessively moist soil, which lowers microbial activity, air permeability, and the rate at which oxygen is transferred. Furthermore, treating low-permeable soils with bioslurping is inappropriate (Godheja et al., 2019). Additionally, bioreactors and other engineered systems require substantial maintenance and operational control.

9.6.4   GENETIC AND METABOLIC CONSTRAINTS

Microorganisms possess inherent genetic and metabolic constraints that limit their ability to degrade certain pollutants. Genetic engineering holds promise for enhancing microbial capabilities, but it also raises concerns regarding the stability and ecological impact of genetically modified organisms (GMOs) in the environment. Using Recombinant Deoxyribonucleic acid (rDNA) technology for metabolic engineering, it is possible to create a single bacterial stain that combines many breakdown pathways from distinct sources. For effective in situ bioremediation, it offers selective advantages to break down the various contaminants at the target sites (Gupta and Shukla, 2016). Degradation of pollutants often involves complex metabolic pathways requiring the coordinated action of multiple genes. Disruption in any part of the pathway can halt the degradation process. Engineered microorganisms or those naturally evolving to degrade pollutants must maintain genetic stability to function effectively over time. Genetic mutations can reduce their bioremediation capabilities. The treatment of contaminated soil, sewage sludge, surface water, and groundwater has shown tremendous potential for genetically engineered microbes (GEMs) due to their capacity to mineralize a wide range of synthetic chemicals (Sarma and Prasad, 2019). The most researched heavy metal that is harmful and released through a variety of man-made and natural processes is mercury. A report on the Hg reductase enzyme's detoxification of mercury to harmless mercury is available (Mukkata et al., 2015). The operon's (mer) clustered genes serve as the foundation for this system. Various kinds of bacteria, including *Xanthomonas* sp. W17, *Staphylococcus aureus*, *Acinetobacter calcoaceticus,* transposon Tn21 (*Shigella flexneri*), etc., have distinct types of mer operons (Singh et al., 2011). Effective monitoring and assessment of bioremediation progress are essential but challenging. Traditional methods such as chemical analysis of soil and water samples can be time-consuming and expensive. Advanced molecular techniques, including metagenomics and proteomics, offer more detailed insights but require specialized equipment and expertise. The ability of bioreactor-based bioremediation to efficiently establish controlled bioaugmentation, nutrient addition, greater accessibility of pollutants, and mass flow (contact between bacteria and pollution) are some of the limiting elements in this process. Any barrier that is not properly handled or perhaps maintained in an ideal state may become a limiting factor, that can minimize microbial growth and make an assessment of bioremediation more difficult due to a few bioprocess limits. In conclusion, different bioreactors will likely have varied effects on pollution; therefore, it's critical to know which approach is the best. Furthermore, there are a few reasons why bioremediation based upon bioreactor is not a widely recognized full-scale method. Because bioreactors use an ex situ method, handling large amounts of polluted soil or other chemicals may necessitate extra work, resources, and safety measures in order to remove the toxins from that particular area (Hussain et al., 2022).

9.6.5 OVERCOMING BARRIERS

Toxic substances have been eliminated from the environment more quickly and effectively in recent years because of the integration of nanomaterials (NMs) with biological processes. The increasing significance of microalgae in altering the role of the bioremediation process makes their application crucial in the current context. Since the current trend in bioremediation goes beyond remediation alone to improve carbon footprint, the best energy recovery can be achieved by extracting many value-added products from the treatment process in order to reduce the cost of treatment costs (Patel et al., 2021). Due to their special qualities, such as negatively charged surfaces, prolonged stagnation in the aqueous phase, large specific surface areas, and high gas transfer efficiency, this indicates effective mass transfer in the gaseous-aqueous phase. Micro and nanobubble technology (MNBT) showed great potential in sewage water, groundwater, and sludge bioremediations. Compared to other gases, air, ozone, and oxygen are used extensively in environmental bioremediation (Xiao et al., 2019). The development of biochar research, particularly in terms of its tailored features, has broader applications in enhancing both inorganic and organic bioremediation. The initial application of biochar was limited to the binding and elimination of pollutants from contaminated environments; however, more recent methods have expanded their scope to include the eventual breakdown of both organic and inorganic contaminants. It is possible to accomplish pollutant degradation in the biochar system with or without the use of bacteria. More focus is placed on newly developed tailoring techniques that use innovatively tailored biochar to remediate a broad range of contaminants (Kumar Patel et al., 2021).

9.7 CONCLUSIONS

Bioremediation, while promising as an eco-friendly and cost-effective method for detoxifying polluted environments, faces significant challenges. The heterogeneity of contaminated sites is one of the environmental barriers that complicates the distribution and efficacy of bioremediating agents. Factors such as temperature, pH, nutrient availability, and the presence of co-contaminants can inhibit microbial activity and biodegradation processes. Moreover, the genetic adaptability of native microbial populations to contaminants is often limited, necessitating the introduction of exogenous strains, which may struggle to survive and compete in foreign ecosystems. The development of effective bioremediation strategies requires advanced techniques for site assessment, monitoring, and control. Current methods for detecting and quantifying pollutants and microbial activity are often inadequate for real-time, in situ applications. Addressing these challenges demands an interdisciplinary approach that integrates environmental science, microbiology, engineering, and technology development. Advancements in molecular biology, such as genomic and metagenomic analyses, can enhance our understanding of microbial diversities as well as the pollutants' interactions. Additionally, innovative technologies, including nanotechnology and bioaugmentation, hold promise for overcoming current limitations. Collaboration among researchers, industry stakeholders, and policymakers is essential to drive research, funding, and regulatory frameworks that support the practical application of bioremediation. Ultimately, overcoming these environmental and technological barriers will be important for realizing the effectiveness of bioremediation in achieving sustainable environmental restoration.

REFERENCES

Agarry, S.E., Ogunleye, O.O., 2012. Box-Behnken design application to study enhanced bioremediation of soil artificially contaminated with spent engine oil using biostimulation strategy. *Int. J. Energy Environ. Eng.* 3, 31. https://doi.org/10.1186/2251-6832-3-31

Akhtar, S., Mohsin, A., Riaz, A., Mohsin, F., 2023. Worldwide efficiency of bioremediation techniques for organic pollutants in soil: a brief review. *Geosfera Indones.* 8, 102. https://doi.org/10.19184/geosi.v8i1.30875

Arora, S., Saxena, S., Sutaria, D., Sethi, J., 2022. Bioremediation: an ecofriendly approach for the treatment of oil spills, in: Das, P., Manna, S., Pandey, J.K. (Eds.), *Advances in Oil-Water Separation*. Unites States-Elsevier, pp. 353–373. https://doi.org/10.1016/B978-0-323-89978-9.00012-4

Ayilara, M.S., Babalola, O.O., 2023. Bioremediation of environmental wastes: the role of microorganisms. *Front. Agron.* 5, 1183691. https://doi.org/10.3389/fagro.2023.1183691

Azubuike, C.C., Chikere, C.B., Okpokwasili, G.C., 2016. Bioremediation techniques–classification based on site of application: principles, advantages, limitations and prospects. *World J. Microbiol. Biotechnol.* 32, 180. https://doi.org/10.1007/s11274-016-2137-x

Bajagain, R., Gautam, P., Jeong, S.-W., 2021. Improved delivery of remedial agents using surface foam spraying with vertical holes into unsaturated diesel-contaminated soil for total petroleum hydrocarbon removal. *Appl. Sci.* 11, 781. https://doi.org/10.3390/app11020781

Bajagain, R., Gautam, P., Jeong, S.-W., 2020. Biodegradation and post-oxidation of fuel-weathered field soil. *Sci. Total Environ.* 734, 139452. https://doi.org/10.1016/j.scitotenv.2020.139452

Bala, S., Garg, D., Thirumalesh, B.V., Sharma, M., Sridhar, K., Inbaraj, B.S., Tripathi, M., 2022. Recent strategies for bioremediation of emerging pollutants: a review for a green and sustainable environment. *Toxics* 10, 484. https://doi.org/10.3390/toxics10080484

Bhakat, K., Chakraborty, A., Islam, E., 2019. Characterization of arsenic oxidation and uranium bioremediation potential of arsenic resistant bacteria isolated from uranium ore. *Environ. Sci. Pollut. Res.* 26, 12907–12919. https://doi.org/10.1007/s11356-019-04827-6

Bodor, A., Petrovszki, P., Erdeiné Kis, Á., Vincze, G.E., Laczi, K., Bounedjoum, N., Szilágyi, Á., Szalontai, B., Feigl, G., Kovács, K.L., Rákhely, G., Perei, K., 2020. Intensification of ex situ bioremediation of soils polluted with used lubricant oils: a comparison of biostimulation and bioaugmentation with a special focus on the type and size of the inoculum. *Int. J. Environ. Res. Public. Health* 17, 4106. https://doi.org/10.3390/ijerph17114106

Chandra, S., Sharma, R., Singh, K., Sharma, A., 2013. Application of bioremediation technology in the environment contaminated with petroleum hydrocarbon. *Ann. Microbiol.* 63, 417–431. https://doi.org/10.1007/s13213-012-0543-3

Chaudhary, D.K., Bajagain, R., Jeong, S.-W., Kim, J., 2019. Biodegradation of diesel oil and n-alkanes (C_{18}, C_{20}, and C_{22}) by a novel strain *Acinetobacter* sp. K-6 in unsaturated soil. *Environ. Eng. Res.* 25, 290–298. https://doi.org/10.4491/eer.2019.119

Chen, W., Kong, Y., Li, J., Sun, Y., Min, J., Hu, X., 2020. Enhanced biodegradation of crude oil by constructed bacterial consortium comprising salt-tolerant petroleum degraders and biosurfactant producers. *Int. Biodeterior. Biodegrad.* 154, 105047. https://doi.org/10.1016/j.ibiod.2020.105047

Chettri, D., Verma, Ashwani Kumar, Verma, Anil Kumar, 2024. Bioaugmentation: an approach to biological treatment of pollutants. *Biodegradation* 35, 117–135. https://doi.org/10.1007/s10532-023-10050-5

Chugh, M., Kumar, L., Shah, M.P., Bharadvaja, N., 2022. Algal Bioremediation of heavy metals: An insight into removal mechanisms, recovery of by-products, challenges, and future opportunities. *Energy Nexus* 7, 100129. https://doi.org/10.1016/j.nexus.2022.100129

Cristorean, C., Micle, V., Sur, I.M., 2016. A critical analysis of ex-situ bioremediation technologies of hydrocarbon polluted soils. *ECOTERRA J. Environ. Res. Prot.* 13, 17–29.

Davoodi, S.M., Miri, S., Taheran, M., Brar, S.K., Galvez-Cloutier, R., Martel, R., 2020. Bioremediation of unconventional oil contaminated ecosystems under natural and assisted conditions: a review. *Environ. Sci. Technol.* 54, 2054–2067. https://doi.org/10.1021/acs.est.9b00906

Dehnavi, S.M., Ebrahimipour, G., 2022. Comparative remediation rate of biostimulation, bioaugmentation, and phytoremediation in hydrocarbon contaminants. *Int. J. Environ. Sci. Technol.* 19, 11561–11586. https://doi.org/10.1007/s13762-022-04343-0

Dell'Anno, F., Brunet, C., Van Zyl, L.J., Trindade, M., Golyshin, P.N., Dell'Anno, A., Ianora, A., Sansone, C., 2020. Degradation of hydrocarbons and heavy metal reduction by marine bacteria in highly contaminated sediments. *Microorganisms* 8, 1402. https://doi.org/10.3390/microorganisms8091402

Duc, H.D., Hung, N.V., Oanh, N.T., 2021. Anaerobic degradation of endosulfans by a mixed culture of Pseudomonas sp. and Staphylococcus sp. *Appl. Biochem. Microbiol.* 57, 327–334. https://doi.org/10.1134/S0003683821030030

Dusengemungu, L., Kasali, G., Gwanama, C., Ouma, K.O., 2020. Recent advances in biosorption of copper and cobalt by filamentous fungi. *Front. Microbiol.* 11, 582016. https://doi.org/10.3389/fmicb.2020.582016

Etuk, I.F., Etuk, B.R., Innocent Oseribho Oboh, 2024. An overview of bioremediation process – mechanisms and factors influencing It. *Int. J. Eng. Mod. Technol.* 10, 1–37.

Fragkou, E., Antoniou, E., Daliakopoulos, I., Manios, T., Theodorakopoulou, M., Kalogerakis, N., 2021. In situ aerobic bioremediation of sediments polluted with petroleum hydrocarbons: a critical review. *J. Mar. Sci. Eng.* 9, 1003. https://doi.org/10.3390/jmse9091003

Fu, W., Xu, M., Sun, K., Hu, L., Cao, W., Dai, C., Jia, Y., 2018. Biodegradation of phenanthrene by endophytic fungus Phomopsis liquidambari in vitro and in vivo. *Chemosphere* 203, 160–169. https://doi.org/10.1016/j.chemosphere.2018.03.164

Gaur, N., Narasimhulu, K., PydiSetty, Y., 2018. Recent advances in the bio-remediation of persistent organic pollutants and its effect on environment. *J. Clean. Prod.* 198, 1602–1631. https://doi.org/10.1016/j.jclepro.2018.07.076

Geetha, N., Bhavya, G., Abhijith, P., Shekhar, R., Dayananda, K., Jogaiah, S., 2021. Insights into nanomycoremediation: Secretomics and mycogenic biopolymer nanocomposites for heavy metal detoxification. *J. Hazard. Mater.* 409, 124541. https://doi.org/10.1016/j.jhazmat.2020.124541

Geng, S., Qin, W., Cao, W., Wang, Y., Ding, A., Zhu, Y., Fan, F., Dou, J., 2022. Pilot-scale bioaugmentation of polycyclic aromatic hydrocarbon (PAH)-contaminated soil using an indigenous bacterial consortium in soil-slurry bioreactors. *Chemosphere* 287, 132183. https://doi.org/10.1016/j.chemosphere.2021.132183

Ghosh, D., Ghorai, P., Sarkar, S., Maiti, K.S., Hansda, S.R., Das, P., 2023. Microbial assemblage for solid waste bioremediation and valorization with an essence of bioengineering. *Environ. Sci. Pollut. Res.* 30, 16797–16816. https://doi.org/10.1007/s11356-022-24849-x

Godheja, J., Modi, D.R., Kolla, V., Pereira, A.M., Bajpai, R., Mishra, M., Sharma, S.V., Sinha, K., Shekhar, S.K., 2019. Environmental remediation: microbial and nonmicrobial prospects, in: Singh, D. P., Gupta, V. K., Prabha, R. (Eds.), *Microbial Interventions in Agriculture and Environment*. Springer Singapore, Singapore, pp. 379–409. https://doi.org/10.1007/978-981-13-8383-0_13

Gupta, S.K., Shukla, P., 2016. Advanced technologies for improved expression of recombinant proteins in bacteria: perspectives and applications. *Crit. Rev. Biotechnol.* 36, 1089–1098. https://doi.org/10.3109/07388551.2015.1084264

Hussain, A., Rehman, F., Rafeeq, H., Waqas, M., Asghar, A., Afsheen, N., Rahdar, A., Bilal, M., Iqbal, H.M.N., 2022. In-situ, ex-situ, and nano-remediation strategies to treat polluted soil, water, and air – a review. *Chemosphere* 289, 133252. https://doi.org/10.1016/j.chemosphere.2021.133252

Jabbar, N.M., Alardhi, S.M., Mohammed, A.K., Salih, I.K., Albayati, T.M., 2022. Challenges in the implementation of bioremediation processes in petroleum-contaminated soils: a review. *Environ. Nanotechnol. Monit. Manag.* 18, 100694. https://doi.org/10.1016/j.enmm.2022.100694

Janssen, D.B., Stucki, G., 2020. Perspectives of genetically engineered microbes for groundwater bioremediation. *Environ. Sci. Process. Impacts* 22, 487–499. https://doi.org/10.1039/C9EM00601J

Jeyakumar, P., Debnath, C., Vijayaraghavan, R., Muthuraj, M., 2022. Trends in bioremediation of heavy metal contaminations. *Environ. Eng. Res.* 28, 220631. https://doi.org/10.4491/eer.2021.631

Kästner, M., Miltner, A., 2016. Application of compost for effective bioremediation of organic contaminants and pollutants in soil. *Appl. Microbiol. Biotechnol.* 100, 3433–3449. https://doi.org/10.1007/s00253-016-7378-y

Kharangate-Lad, A., D'Souza, N.C., 2021. Current approaches in bioremediation of toxic contaminants by application of microbial cells; biosurfactants and bioemulsifiers of microbial origin, in: Kumar, V., Prasad, R., Kumar, M. (Eds.), *Rhizobiont in Bioremediation of Hazardous Waste*. Springer Singapore, Singapore, pp. 217–263. https://doi.org/10.1007/978-981-16-0602-1_11

Kumar Patel, A., Tseng, Y.-S., Rani Singhania, R., Chen, C.-W., Chang, J.-S., Di Dong, C., 2021. Novel application of microalgae platform for biodesalination process: a review. *Bioresour. Technol.* 337, 125343. https://doi.org/10.1016/j.biortech.2021.125343

Lamichhane, S., Bal Krishna, K.C., Sarukkalige, R., 2016. Polycyclic aromatic hydrocarbons (PAHs) removal by sorption: a review. *Chemosphere* 148, 336–353. https://doi.org/10.1016/j.chemosphere.2016.01.036

Larik, I.A., Qazi, M.A., Kanhar, A.R., Mangi, S., Ahmed, S., Jamali, M.R., Kanhar, N.A., 2016. Biodegradation of petrochemical hydrocarbons using an efficient bacterial consortium: A2457. *Arab. J. Sci. Eng.* 41, 2077–2086. https://doi.org/10.1007/s13369-015-1851-2

Li, N., Xia, Q., Niu, M., Ping, Q., Xiao, H., 2018. Immobilizing laccase on different species wood biochar to remove the chlorinated biphenyl in wastewater. *Sci. Rep.* 8, 13947. https://doi.org/10.1038/s41598-018-32013-0

Liu, Z., Shao, B., Zeng, G., Chen, M., Li, Z., Liu, Yujie, Jiang, Y., Zhong, H., Liu, Yang, Yan, M., 2018. Effects of rhamnolipids on the removal of 2,4,2,4-tetrabrominated biphenyl ether (BDE-47) by Phanerochaete chrysosporium analyzed with a combined approach of experiments and molecular docking. *Chemosphere* 210, 922–930. https://doi.org/10.1016/j.chemosphere.2018.07.114

Lopes, P.R.M., Cruz, V.H., De Menezes, A.B., Gadanhoto, B.P., Moreira, B.R.D.A., Mendes, C.R., Mazzeo, D.E.C., Dilarri, G., Montagnolli, R.N., 2022. Microbial bioremediation of pesticides in agricultural soils: an integrative review on natural attenuation, bioaugmentation and biostimulation. *Rev. Environ. Sci. Biotechnol.* 21, 851–876. https://doi.org/10.1007/s11157-022-09637-w

Lukić, B., Panico, A., Huguenot, D., Fabbricino, M., Van Hullebusch, E.D., Esposito, G., 2017. A review on the efficiency of landfarming integrated with composting as a soil remediation treatment. *Environ. Technol. Rev.* 6, 94–116. https://doi.org/10.1080/21622515.2017.1310310

Ma, Y.-L., Lu, W., Wan, L.-L., Luo, N., 2015. Elucidation of fluoranthene degradative characteristics in a newly isolated achromobacter xylosoxidans DN002. *Appl. Biochem. Biotechnol.* 175, 1294–1305. https://doi.org/10.1007/s12010-014-1347-7

Mapelli, F., Scoma, A., Michoud, G., Aulenta, F., Boon, N., Borin, S., Kalogerakis, N., Daffonchio, D., 2017. Biotechnologies for marine oil spill cleanup: indissoluble ties with microorganisms. *Trends Biotechnol.* 35, 860–870. https://doi.org/10.1016/j.tibtech.2017.04.003

Mehrotra, T., Dev, S., Banerjee, A., Chatterjee, A., Singh, R., Aggarwal, S., 2021. Use of immobilized bacteria for environmental bioremediation: a review. *J. Environ. Chem. Eng.* 9, 105920. https://doi.org/10.1016/j.jece.2021.105920

Mineo, S., 2023. Groundwater and soil contamination by LNAPL: State of the art and future challenges. *Sci. Total Environ.* 874, 162394. https://doi.org/10.1016/j.scitotenv.2023.162394

Miri, S., Rasooli, A., Brar, S.K., Rouissi, T., Martel, R., 2022. Biodegradation of p-xylene—a comparison of three psychrophilic Pseudomonas strains through the lens of gene expression. *Environ. Sci. Pollut. Res.* 29, 21465–21479. https://doi.org/10.1007/s11356-021-17387-5

Moreira, V.R., Lebron, Y.A.R., Lange, L.C., Santos, L.V.S., 2019. Simultaneous biosorption of Cd(II), Ni(II) and Pb(II) onto a brown macroalgae Fucus vesiculosus: Mono- and multi-component isotherms, kinetics and thermodynamics. *J. Environ. Manage.* 251, 109587. https://doi.org/10.1016/j.jenvman.2019.109587

Mousavi, S.M., Hashemi, S.A., Iman Moezzi, S.M., Ravan, N., Gholami, A., Lai, C.W., Chiang, W.-H., Omidifar, N., Yousefi, K., Behbudi, G., 2021. Recent advances in enzymes for the bioremediation of pollutants. *Biochem. Res. Int.* 2021, 1–12. https://doi.org/10.1155/2021/5599204

Mukkata, K., Kantachote, D., Wittayaweerasak, B., Techkarnjanaruk, S., Mallavarapu, M., Naidu, R., 2015. Distribution of mercury in shrimp ponds and volatilization of Hg by isolated resistant purple nonsulfur bacteria. *Water. Air. Soil Pollut.* 226, 148. https://doi.org/10.1007/s11270-015-2418-2

Muley, E., Chavan, A., Naphade, S., 2023. Bioremediation: a sustainable tool for environmental wellbeing. *Int. J. Ecol. Environ. Sci.* 5, 15–18.

Muttaleb, W.H., Ali, Z.H., 2022. Bioremediation an eco-friendly method for administration of environmental contaminants. *Minar Int. J. Appl. Sci. Technol.* 4. https://doi.org/10.47832/2717-8234.11.3

Nag, M., Lahiri, D., Ghosh, S., Sarkar, T., Pati, S., Das, A.P., Ram, D.K., Bhattacharya, D., Ray, R.R., 2024. Application of microorganisms in biotransformation and bioremediation of environmental contaminant: a review. *Geomicrobiol. J.* 41, 374–391. https://doi.org/10.1080/01490451.2023.2261443

Nanda, M., Kumar, V., Sharma, D.K., 2019. Multimetal tolerance mechanisms in bacteria: The resistance strategies acquired by bacteria that can be exploited to 'clean-up' heavy metal contaminants from water. *Aquat. Toxicol.* 212, 1–10. https://doi.org/10.1016/j. aquatox.2019.04.011

Narayan Chadar, S., 2018. Composting as an eco-friendly method to recycle organic waste. *Prog. Petrochem. Sci.* 2. https://doi.org/10.31031/PPS.2018.02.000548

Naseri, M., Barabadi, A., Barabady, J., 2014. Bioremediation treatment of hydrocarbon-contaminated Arctic soils: influencing parameters. *Environ. Sci. Pollut. Res.* 21, 11250–11265. https://doi.org/10.1007/s11356-014-3122-2

Nayak, P., Solanki, H., 2022. Impact of agriculture on environment and bioremediation techniques for improvisation of contaminated site. *Int. Assoc. Biol. Comput. Dig.* 1, 163–174. https://doi.org/10.56588/iabcd.v1i1.31

Pande, Veni, Pandey, S.C., Sati, D., Pande, Veena, Samant, M., 2020. Bioremediation: an emerging effective approach towards environment restoration. *Environ. Sustain.* 3, 91–103. https://doi.org/10.1007/s42398-020-00099-w

Patel, A.K., Singhania, R.R., Albarico, F.P.J.B., Pandey, A., Chen, C.-W., Dong, C.-D., 2022. Organic wastes bioremediation and its changing prospects. *Sci. Total Environ.* 824, 153889. https://doi.org/10.1016/j.scitotenv.2022.153889

Patel, A.K., Singhania, R.R., Chen, C.-W., Tseng, Y.-S., Kuo, C.-H., Wu, C.-H., Dong, C.D., 2021. Advances in micro- and nano bubbles technology for application in biochemical processes. *Environ. Technol. Innov.* 23, 101729. https://doi.org/10.1016/j. eti.2021.101729

Polyak, Y.M., Bakina, L.G., Chugunova, M.V., Mayachkina, N.V., Gerasimov, A.O., Bure, V.M., 2018. Effect of remediation strategies on biological activity of oil-contaminated soil – a field study. *Int. Biodeterior. Biodegrad.* 126, 57–68. https://doi.org/10.1016/j. ibiod.2017.10.004

Rabani, M.S., Sharma, R., Singh, R., Gupta, M.K., 2022. Characterization and identification of naphthalene degrading bacteria isolated from petroleum contaminated sites and their possible use in bioremediation. *Polycycl. Aromat. Compd.* 42, 978–989. https://doi.org/ 10.1080/10406638.2020.1759663

Saavedra, R., Muñoz, R., Taboada, M.E., Vega, M., Bolado, S., 2018. Comparative uptake study of arsenic, boron, copper, manganese and zinc from water by different green microalgae. *Bioresour. Technol.* 263, 49–57. https://doi.org/10.1016/j.biortech.2018.04.101

Sarma, H., Prasad, M.N.V., 2019. Metabolic engineering of rhizobacteria associated with plants for remediation of toxic metals and metalloids, in: Prasad, M.N.V. (Ed.), *Transgenic Plant Technology for Remediation of Toxic Metals and Metalloids.* United States, Academic Press, Elsevier, pp. 299–318. https://doi.org/10.1016/B978-0-12-814389-6.00014-6

Sayed, K., Baloo, L., Sharma, N.K., 2021. Bioremediation of total petroleum hydrocarbons (TPH) by bioaugmentation and biostimulation in water with floating oil spill containment booms as bioreactor basin. *Int. J. Environ. Res. Public. Health* 18, 2226. https:// doi.org/10.3390/ijerph18052226

Shah, H., Jain, S., 2020. Bioremediation: An Approach for Environmental Pollutants Detoxification, in: Kashyap, B. K., Solanki, M. K., Kamboj, D. V., Pandey, A. K. (Eds.), *Waste to Energy: Prospects and Applications.* Springer Singapore, Singapore, pp. 121–142. https://doi.org/10.1007/978-981-33-4347-4_6

Sharma, B., Dangi, A.K., Shukla, P., 2018. Contemporary enzyme based technologies for bioremediation: A review. *J. Environ. Manage.* 210, 10–22. https://doi.org/10.1016/j. jenvman.2017.12.075

Singh, J.S., Abhilash, P.C., Singh, H.B., Singh, R.P., Singh, D.P., 2011. Genetically engineered bacteria: An emerging tool for environmental remediation and future research perspectives. *Gene* 480, 1–9. https://doi.org/10.1016/j.gene.2011.03.001

Sutherland, D.L., Ralph, P.J., 2019. Microalgal bioremediation of emerging contaminants - Opportunities and challenges. *Water Res.* 164, 114921. https://doi.org/10.1016/j.watres.2019.114921

Svinterikos, E., Zuburtikudis, I., Al-Marzouqi, M., 2019. Carbon nanomaterials for the adsorptive desulfurization of fuels. *J. Nanotechnol.* 2019, 1–13. https://doi.org/10.1155/2019/2809867

Tyagi, M., Da Fonseca, M.M.R., De Carvalho, C.C.C.R., 2011. Bioaugmentation and biostimulation strategies to improve the effectiveness of bioremediation processes. *Biodegradation* 22, 231–241. https://doi.org/10.1007/s10532-010-9394-4

Wang, L., Rinklebe, J., Tack, F.M.G., Hou, D., 2021. A review of green remediation strategies for heavy metal contaminated soil. *Soil Use Manage.* 37, 936–963. https://doi.org/10.1111/sum.12717

Wu, M., Wu, J., Zhang, X., Ye, X., 2019. Effect of bioaugmentation and biostimulation on hydrocarbon degradation and microbial community composition in petroleum-contaminated loessal soil. *Chemosphere* 237, 124456. https://doi.org/10.1016/j.chemosphere.2019.124456

Xiao, Z., Aftab, T.B., Li, D., 2019. Applications of micro–nano bubble technology in environmental pollution control. *Micro Nano Lett.* 14, 782–787. https://doi.org/10.1049/mnl.2018.5710

Yap, H.S., Zakaria, N.N., Zulkharnain, A., Sabri, S., Gomez-Fuentes, C., Ahmad, S.A., 2021. Bibliometric analysis of hydrocarbon bioremediation in cold regions and a review on enhanced soil bioremediation. *Biology* 10, 354. https://doi.org/10.3390/biology10050354

Zouboulis, A.I., Moussas, P.A., 2019. Groundwater and soil pollution: bioremediation, in: Nriagu, J. (Ed.), *Encyclopedia of Environmental Health*. Oxford: Elsevier, pp. 369–381. https://doi.org/10.1016/B978-0-444-52272-6.00035-0

10 Bioremediation Strategies for Petroleum Hydrocarbons, Heavy Metals, and Pesticides

Naseeb, Sushmita Gandash,
Pradeep Kumar, and Adarsh Kumar Shukla

10.1 INTRODUCTION

The breakdown of petroleum hydrocarbons (PHCs), a class of complex organic chemicals made up of nonhydrocarbon molecules such as naphthenic acid, phenol, thiol, and asphaltene, can be accomplished creatively and sustainably through the use of bioremediation. Because of the high petroleum production required to keep up with the rapid expansion of the industrial, agricultural, and municipal sectors, these compounds constitute substantial pollutants. With 100.90 million barrels of petroleum products predicted to be consumed globally each day, leaks from drilling, exploration, storage, transportation, and refining operations would likely increase environmental contamination.

PHC leaks cause soil contamination that alters the carbon/nitrogen ratio, lowers microbial activity, and depletes the soil's oxygen reserves, all of which pose serious concerns to ecosystems and human health. This emphasizes how crucial it is to develop efficient soil remediation techniques in order to maintain soil ecosystems. PHC-contaminated soils have been treated using a variety of remediation strategies, such as physical, chemical, and biological techniques. Among these, bioremediation is the most affordable and environmentally friendly option.

By using the metabolic powers of microbes and plants, bioremediation converts PHCs into innocuous byproducts, including CO_2, H_2O, and simple organic molecules. Environmental elements, including temperature, pH, and oxygen availability, have an impact on this process. The capacity of microorganisms to break down hydrocarbons makes them essential to this process, including fungi, bacteria, and algae. Microalgae such as *Chlorella* and *Scenedesmus*, fungal strains like *Aspergillus* and *Penicillium*, and bacterial strains like *Pseudomonas* and *Rhodococcus* have demonstrated considerable potential in PHC decomposition.

Optimizing the environment for microbial activity, which includes making sure there is an appropriate supply of nutrients and oxygen, is what makes bioremediation efficient. The introduction of microorganisms that break down pollutants or the

DOI: 10.1201/9781003407317-10

addition of nutrients to promote microbial development are examples of techniques like bioaugmentation and biostimulation that improve the degradation process. These techniques have produced encouraging outcomes in a number of case studies, highlighting the potential of bioremediation to reduce pollution caused by petroleum hydrocarbons.

Pesticides are primarily used to target different pests, including weeds, fungi, insects, and rodents, in order to increase crop output. Unfortunately, widespread pesticide use—especially in agricultural settings—has contaminated soil and groundwater, endangering human and environmental health. For example, crops like rice and cotton use a lot of pesticides, which means that every year, some 4.6 million tons of toxic chemicals are released into the environment worldwide.

Because pesticides are used so extensively in agriculture, environmental pollution from these chemicals is a serious concern. Pesticides are categorized according to their chemical structure and include pyrethroids, carbamates, pyrethrin, organochlorine, and organophosphorus. The toxicity and environmental persistence of each type vary; pyrethrin and pyrethroids are more biodegradable and comparatively less dangerous to mammals.

There are significant dangers associated with traditional techniques of cleaning up pesticide-contaminated sites, such as chemical treatments and excavation. As a result, bioremediation—the process of breaking down pesticides using microorganisms—has become a viable substitute. This method converts poisonous pesticides into less dangerous compounds by making use of the natural metabolic activities of bacteria and fungi. The two types of bioremediation procedures are in situ methods, which take place at the site of contamination, and ex situ methods, which entail moving the contaminants to another location for treatment.

In situ bioremediation includes techniques like bioaugmentation, biosparging, and bioventing, while ex situ approaches entail procedures like composting and slurry bioreactors to increase microbial activity and pesticide breakdown. These techniques focus on adjusting environmental factors like pH, temperature, oxygen levels, and nutrition availability. Case studies demonstrating successful pesticide removal from contaminated soils and water bodies demonstrate the efficacy of bioremediation. For example, decomposing particular pesticide isomers and agricultural chemicals has been successfully achieved using land farming and bioaugmentation, respectively.

Even though bioremediation shows promise, there are still issues, especially in underdeveloped nations with low resources where the effects of prolonged pesticide exposure are severe. Governments, environmental protection organizations, farmers, and manufacturers must work together to establish strict laws, encourage the safe use of pesticides, and adopt integrated pest management techniques in order to address these issues. It is feasible to lessen the need for dangerous pesticides and lessen their negative impacts on the environment and public health by taking the right steps.

In conclusion, bioremediation is an effective and sustainable way to lessen the negative impacts that PHC contamination has on the ecosystems of soil and water. Using the inherent powers of plants and microbes, this strategy provides a workable way to preserve ecological equilibrium while shielding the general public from the risks posed by petroleum hydrocarbons.

10.2   BIOREMEDIATION OF PETROLEUM HYDROCARBONS

Petroleum hydrocarbons are made up of nonhydrocarbon molecules including naphthenic acid, phenol, thiol, metalloporphyrin, asphaltene, and heterocyclic nitrogen and sulfur compounds, as well as short-chain hydrocarbons like paraffin, alicyclic, and aromatic compounds.

The remarkable rate of population growth worldwide combined with the quick expansion of industrial, agricultural, and municipal activities has created a strong demand for petroleum production (Singha and Pandey, 2021). It is anticipated that the daily demand for petroleum products worldwide will reach 100.90 million barrels (Gaur et al., 2021). Drilling, exploration, storage, transportation, processing, and refining are among the petroleum industry activities that contribute most to the hydrocarbon spillage that results in terrestrial oil pollution, which typically contains hydrophobic chemicals (Chettri et al., 2021). PHCs leak into the environment and can cause major health and environmental issues for humans as well as be a major hazard to ecosystems (Guarino et al., 2017). It is predicted that 0.6 million tons of crude oil leak into the environment each year (Kvenvolden et al., 2003). PHCs are mostly made of carbon and hydrogen, and when they contaminate soil, the proportion of C/N fraction at the spill site is altered. This results in a nitrogen shortage and reduces microbial activity and soil fertility. Furthermore, enormous amounts of organic compounds that break down in the top layer of the soil deplete oxygen reserves and slow the rate at which oxygen diffuses into deeper layers (Dindar et al., 2013). Consequently, the cleanup of PHC-contaminated soils is currently a major undertaking, and the development of sustainable and efficient methods to maintain soil ecosystems is essential (Cui et al., 2020).

Numerous remediation technologies have been demonstrated and implemented to address petroleum-contaminated soils. A real polluted site might necessitate the best remediation through a blend of processes for typical circumstances (Dindar et al., 2013). Physical, chemical, and biological approaches are among the most recent PHC remediation techniques (Shi et al., 2022). Physical and chemical pollutant detoxification technologies are frequently costly and carry the risk of additional air or water pollution. Biological approaches, because of the diverse metabolic capabilities of microorganisms and plants, are advancing techniques for the breakdown of various environmental pollutants containing petroleum products (Singha and Pandey, 2021). Bioremediation is an efficient, cost-effective, and environmentally friendly approach that has been proven to be effective for mineralizing organic contaminants in wastewater and soil environments (Mekonnen et al., 2021). Researchers have been particularly interested in bioremediation of PHC-contaminated soil because of its environmentally friendly nature. In the process of restoring polluted soils, microbes are essential to the biotransformation of complex pollutant combinations. Numerous plant species, as well as bacteria, fungi, and some algae, have been proven to be able to fully mineralize polycyclic aromatic hydrocarbons (PAHs) into CO_2, H_2O, inorganic chemicals, cell proteins, and simple organic compounds that are found in soil and water. However, environmental parameters such as temperature, pH, metabolite inhibition, bioavailability, biochemistry, energy sources, bioavailability, bioactivity, and availability of oxygen and nutrients all affect the biological activity of microbes (Kolukirik et al., 2011). Bacteria and other microorganisms are used as bioremediation agents to change dangerous organic contaminants into safe substances. This procedure has the potential to be economical, ecologically benign, and efficient. Microbial oil extraction and pollution removal are made possible by the substantial enhancement of catalytic capabilities by microbes. The most frequent method for preventing petroleum hydrocarbon pollution of marine and agricultural habitats is microbial bioremediation. Due to their genetic adaptations to petroleum-contaminated settings, microbes such as bacteria, fungi, and algae play a crucial role in the degradation of petroleum pollutants. Because of their effectiveness in degrading hydrocarbons, bacterial strains like *Pseudomonas* and *Rhodococcus*, as well as fungal strains like *Aspergillus* and *Penicillium*, have been extensively utilized in biodegradation. Furthermore, in aquatic environments, microalgae like *Chlorella* and *Scenedesmus* have the capacity to absorb and metabolize hydrocarbons, providing flexible bioremediation (Alaidaroos et al., 2023).

10.2.1 Aerobic Degradation

Hydrocarbons break down quickly in an aerobic environment. Algae, fungi, and bacteria can all degrade hydrocarbons aerobically (Haritash et al., 2009). The most easily broken-down types of hydrocarbons are usually alkenes (those with two bonds) and short-chain alkanes (those with one bond), which are followed by branched alkanes (those with side chains) and aromatics (those with a stable ring structure) (Xue et al., 2015). Degradation rates, however, vary with environmental factors and get less complex as hydrocarbons get more complex. Because the hydrocarbon composition varies depending on the age of the spill and the source of the petroleum, reported degradation rates can differ significantly. For instance, after 28 days, a compound breakdown can range from 5% to 30%, while adding nitrogen can cause up to 100% degradation. Fungal species have been shown to exhibit degradation rates ranging from approximately 30% to 100% in less than 28 days (Zafra et al., 2015). As was previously said, oxygen delivery is the main element that limits the pace of aerobic biodegradation. The capacity of oxygen to flow or diffuse throughout the site environment and the pace at which microbes absorb it determine the availability of oxygen. Degradation rates can be several orders of magnitude higher when oxygen is added than when it occurs naturally. Algae, fungi, and bacteria may all biodegrade hydrocarbons in an aerobic manner. Aromatic hydrocarbons are widely distributed and exist in a variety of chemical forms. Bacteria in particular break down these structures. The capacity of aromatic hydrocarbons to change through peripheral reactions is the fundamental idea behind their biodegradation. Additionally, the TCA Tricarboxylic Acid cycle (TCA) cycle is funneled into a small range of core intermediates, which are then exposed to ring cleavage (Sun et al., 2019). The rate of biodegradation is contingent upon multiple ecological factors. This falls off as hydrocarbon complexity rises. When a nitrogen source is added, biodegradation efficiency is seen since nitrogen is known to meet some microbes' nutritional needs. Petroleum hydrocarbons typically contain three types of molecules: alkanes, cyclo-alkanes, and hydrocarbon mono-aromatics known as benzene, toluene, ethylbenzene, and xylene (BTEX), which stands for benzene, toluene, ethylbenzene, and xylene isomers. In addition to polluting the environment, BTEX chemicals' toxicity poses a risk to human health (Kamani et al., 2023). Two enzyme systems, namely dioxygenases and monooxygenases, provide a metabolic pathway for the aerobic breakdown of the BTEX chemicals. Attacked by the monooxygenase enzyme, the aromatic ring structure's ethyl or methyl substituents are converted into matching substituted phenylglyoxal or pyrocatechol by a series of oxidation processes.

10.2.2 ALKANES AND ALKENES

Alkanes with more carbons are less volatile than those with fewer carbons, which are more likely to volatilize. All things considered, alkanes and alkenes are the most easily broken-down hydrocarbons, with the exception of cyclic alkanes (alkanes having a ring structure), with reports of alkanes with up to 44 carbons breaking down. According to Chikere et al. (2021), the addition of molecular oxygen causes the degradation of both alkanes and alkenes. The first stage of degradation and the availability of oxygen are rate-limiting factors. The process of aerobic alkane breakdown begins when enzymes known as oxygenase introduce molecular oxygen to hydrocarbon molecules (Chikere et al., 2011). This results in the formation of alcohols, which are then further oxidized to fatty acids, which are then digested to produce acetyl-coenzyme A (CoA), CO_2, and H_2O. Numerous oxidases, including mono- and dioxygenase, have a broad spectrum of substrates and act readily on a variety of hydrocarbons (Abbasian et al., 2015).

10.2.3 AROMATIC HYDROCARBONS

Due to their higher toxicity, aromatic hydrocarbons are often more difficult to break down than shorter alkanes and alkenes; nevertheless, many bacteria and fungi can easily break them down aerobically (see also PAHs). Because of their greater hydrophobicity and sorption capability, molecules with more rings and larger molecular sizes are less degradable (Chikere et al., 2011). BTEX compounds have a median primary degradation rate of 0.05–0.2 per day. For every 1 mg/L of BTEX, 3.1 mg/L of dissolved oxygen (DO) is needed for BTEX oxidation. Biodegradation slows down when DO is less than 2 mg/L (Lawrence et al., 2006). The addition of O_2 by mono- and dioxygenases initiates the overall pathway for the breakdown of aromatic compounds (Yadav et al., 2020). Important intermediate chemicals such as benzyl alcohol, protocatechuate, phenol or catechol, and gentisate are produced as a result (Gogoi et al., 2022). Following ring breakage by a range of oxygenases, these intermediates become carboxylic acids (Denkler et al., 2024). After that, degradation proceeds to produce succinyl- and acetyl-CoAs (Zeaiter et al., 2024), which are metabolized centrally. Despite the fact that many of the processes are comparable to those in bacteria, non-specific extracellular oxidizing enzymes that produce radical intermediates are responsible for the destruction of bacteria (Mawardi et al., 2023).

Because of the less favorable reaction energies with alternative electron acceptors, anaerobic processes generally impact the fate of hydrocarbons in the environment more than aerobic ones. However, both facultative and obligately anaerobic bacteria and archaea are known to degrade hydrocarbons without oxygen; these microorganisms develop readily at hydrocarbon-impacted sites because they consume oxygen quickly. The first steps in anoxic hydrocarbon degradation, which involve adding an oxidized functional group to activate the molecule, are usually rate-limiting. Anaerobic hydrocarbon degraders can double in length from a few days to several months (Kolukirik et al., 2011). In the absence of oxygen, many different forms of hydrocarbons completely degrade, notwithstanding their sluggish development rates. For instance, it has been observed that the breakdown of molecules containing many aromatic rings, or PAHs, takes place in 90 days, but the degradation of benzene takes place over 120 weeks (Li et al., 2024). Linear alkanes have reportedly decomposed under methanogenic conditions in less than 200 days (Laban et al., 2014).

In addition to oxygen, anaerobic microorganisms respire using alternative terminal electron acceptors, such as nitrate, sulfate, carbon dioxide, oxidized metals, or even some organic compounds (Pandit, 2022). Microbes at a contaminated site typically use oxygen, nitrate, ferric iron, sulfate, and H_2 as electron acceptors in that order as a function of diminishing reduction potential (Costa et al., 2024). It has been shown in a few instances that particular species of denitrifying or sulfate-reducing bacteria may fully metabolize some hydrocarbons into CO_2 and water. Nevertheless, syntropy—a process whereby the breakdown of a substrate by one microbe depends on the activity of another bacterium that maintains intermediate products like format and H_2 at low concentrations—is more frequently used in the anaerobic degradation of hydrocarbons. Reactions that are otherwise thermodynamically unfavorable are driven by low product concentrations. Anaerobic environments promote syntropy more frequently because oxygen is a more energetically advantageous terminal electron acceptor (Zindel et al., 1988). Since methanogens (archaea that make methane) can only metabolize simple substrates like acetate and hydrogen, syntrophic activities are absolutely necessary for the complete breakdown of methane and carbon dioxide. Depending on the circumstances and substrates that are accessible, a given habitat may have multiple syntrophic interactions. Hydrocarbons are broken down by primary degraders like *Peptococcaceae* and *Clostridium* into intermediates like H_2 and acetate, which are then absorbed by methanogens in methanogenic conditions where all other electron acceptors have been used (Zhang et al., 2020). As shown in Figure 10.1, the acetate and H_2 are consumed in reactions 1, 2, and 3, keeping the fermentation reaction energetically favorable. When external electron acceptors (e.g., nitrate, iron, or sulfate) are no longer available, methanogens consume acetate and hydrogen.

Diverse techniques are employed by anaerobic microorganisms to activate hydrocarbons without the need for molecular oxygen (O_2). Each tactic, which can be used with both aromatic and aliphatic molecules, is covered in detail below. To make a molecule more reactive to subsequent transformation into more frequent intermediates (usually fatty acids and other carboxylic acids) that can enter key metabolic pathways, the standard approach is to introduce a more oxidized group into the molecule. The majority of aromatic chemicals undergo activation and are directed toward the benzoate central anaerobic intermediate, or more precisely, the benzoyl-CoA thioester derivative of CoA.

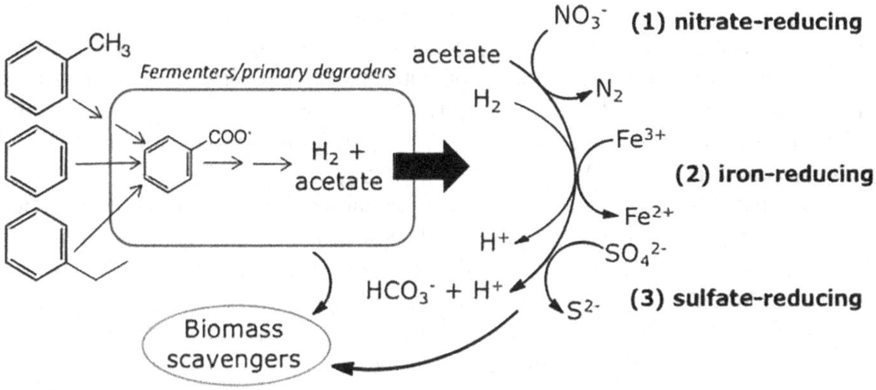

FIGURE 10.1 Conceptual model for syntrophic anaerobic degradation of benzene and alkylbenzenes (Conrad, 2020).

10.2.4 BIOAUGMENTATION

By adding pollutant-degrading microorganisms and increasing their variety, bioaugmentation extends the breakdown of complex pollutants (Adams et al., 2020). In order to expedite the elimination of undesirable substances, this approach entails adding native microbes or genetically modified microorganisms to contaminated areas under ideal conditions (Mrozik et al., 2010). Numerous microorganisms, including fungi, bacteria, and yeast, are capable of breaking down petroleum hydrocarbons. Approximately 200 species of bacteria, fungi, and yeast are recognized as hydrocarbon degraders, and they can break down both simple hydrocarbons like methane and complex hydrocarbons with more than 40 carbons. The decontamination rate is one of the main factors that shortens the time it takes to start the bioremediation process (Bala et al., 2022). For the biodegradation of petroleum hydrocarbon oil (14,000 mg/kg), biostimulation (added with rhamnolipid, low-level and high-level nutrients), bioaugmentation [introduction of kitchen waste (KW) and selected microbial consortiums], and biostimulation plus bioaugmentation (added both bacteria consortia and rhamnolipid) (Ossai et al., 2020), six bio pile batches with different remediation strategies were examined. After 140 days of operation, the highest petroleum hydrocarbons—more than 80%—come from low nutrient level (NEL) and KW batches. According to hydrocarbon analysis, the bioaugmentation strategy is more advantageous and successful at eliminating aromatic chemicals (64% and 68%); KW and NEL procedures, on the other hand, might favorably eradicate the inferior components, 11% and 21%, respectively (Martelletti 2015).

10.2.5 BIOSTIMULATION

The most popular technique for increasing the rate at which PHC-polluted soil detoxifies is biostimulation, which involves adding nutrients such as carbon, nitrogen, phosphorus, and oxygen (MartínezÁlvarez et al., 2017). However, since it could lead to a nutrient imbalance in the microbiota, the concentration of additional nutrients needs to be regulated. By providing nutrients to heavily contaminated areas, biostimulation of PHCs encourages bacteria to break down hazardous and toxic pollutants quickly (Guarino et al., 2017). The most common nutrients that encourage microbial growth (phosphate, salts, and nitrate) and highly effective Hydrocarbon (HC)-degrading microorganisms can be added in situ to speed up biostimulation techniques for PHC mineralization in the soil (Mekonnen et al., 2024).

Additionally, Turgay et al. looked into how hydrocarbon breakdown in oil-contaminated soil was impacted by bioaugmentation and biostimulation (powdered gyttja and its humic-fulvic extract—HFA) (Dindar et al., 2015). Following a 60-day incubation period, the results showed that biostimulation is a more effective technique, yielding 51%–56% oil remodel, whereas bioaugmentation, utilizing a commercial bioremediation product, produced 50% oil remediation. In the control therapy, which produced a 46% performance, only nutrition was used. A researcher discovered that adding beneficial organic compounds, such as humic compounds, to contaminated soil can be just as beneficial as bioaugmentation. By providing the hydrocarbon-degrading microorganisms with these extra nutrients at the proper concentration, they can achieve a maximum growth rate of pollutant absorption (Trejos Delgado et al., 2020).

10.2.6 ANAEROBIC DEGRADATION

10.2.6.1 Monoaromatic Compounds

Monoaromatic hydrocarbons, also referred to as BTEX (Figure 10.1), are extremely volatile compounds that are frequently found in petrol, as shown in Figure 10.2. Owing to their considerable toxicity and solubility, they pose a serious threat to human health and are mostly released into the environment during the production, handling, and storage of petroleum fuels and petrol. A wide range of industrial effluents, such as those used in the production of metal, paint and textiles, wood processing, chemicals, and tobacco products, also produce significant emissions when they are discharged into waterways (Usman and colleagues, 2020).

Benzene	C_6H_6
Toluene	C_7H_8
Ethylbenzene	C_8H_{10}
Xylene	C_8H_{10}

FIGURE 10.2 Molecular formula of the respective BTEX components.

10.2.6.2 Anaerobic Biodegradation of BTEX

The microbial transformation of xylenes in anoxic circumstances was first confirmed in the mid-1980s, while the anaerobic biodegradation of BTEX was long thought to be problematic. Since the 1990s, reports of the biodegradation of aromatic compounds in the absence of oxygen, including benzene, toluene, and ethylbenzene, have been made in addition to xylenes. Aromatic compounds provide electrons to a variety of electron acceptors, including NO 3^-, Fe^{3+}, SO_4^{2-}, and HCO_3^-, during the anaerobic biodegradation of BTEX. Fumarate addition initiates the anaerobic breakdown of ethylbenzene, xylenes, and toluene. Apart from fumarate addition, nitrate-reducing bacteria also create a dehydrogenase that oxidizes ethylbenzene. The exact method for the anaerobic breakdown of benzene is yet unknown. According to Weelink et al. (2010), several anaerobic BTEX degraders have been discovered. The most frequently isolated are those that use nitrate as an electron acceptor, including *Thaueraaromatica* K172 (5), *Azoarcus* sp. T (44), and *Aromatoleumaromaticum* EbN1 (151). Furthermore, microbes that employ sulfate and ferric iron as electron accepters, including *Desulfobaculatoluolica* Tol2 (150) and *Geobactergrbiciae* TACP-2T (36) have also been discovered. Stable isotope probing has demonstrated that *Desulfobacterales* and *Coriobacteriaceae* members participate in the anaerobic breakdown of benzene under methanogenic conditions. Only lately have p-xylene-degrading microorganisms been identified; Higashioka et al. (2012) isolated *Desulfosarcina* sp. PP31 as a degrader under sulfate-reducing conditions. A benzyl succinate synthase (BssA) catalysis is the first step in the anaerobic toluene breakdown process, which involves adding fumarate to toluene. BssA may also catalyze fumarate addition to m-xylene (Yoshikawa et al., 2017).

10.2.7 CASE STUDY OF BIOREMEDIATION

Under ideal circumstances, *Bacillus cereus* was able to produce a potential biosurfactant that demonstrates 90% emulsification activity and a 60% decrease in surface tension in a mineral medium enriched with leftover frying oil. The biosurfactant demonstrated favorable characteristics for use in the bioremediation of seawater tainted with petroleum derivatives, including its capability for dispersion, stability in a range of environmental settings, and efficacy in promoting the biodegradation of motor oil in seawater samples. A recent study showed the potential of the *filamentous ascomycete* fungi, isolated from a polluted site, to biodegrade diesel oil. At the conclusion of the 40-day trial at 25°C, Nazifa et al. demonstrated enhanced degradation of total petroleum hydrocarbons (TPHs) up to 94.78% (Nazifa et al., 2018).

Oil spill biodegradation and agriculture both employ biosurfactants, which Serratia marcescens UCP 1549 is able to produce. Surface active substances known as biosurfactants are more biocompatible and biodegradable than conventional surfactants. *Pseudomonas sp.* sp48, a marine bacteria isolated from the Bahari area of Alexandria, Egypt, demonstrated notable oil degradation of up to 1.5%. It also demonstrated the capacity to use the aliphatic and aromatic hydrocarbon fractions. The Energy and Resources Institute, or TERI, has created the oil zapper, a patented consortium of oil-degrading bacteria that were isolated from various bacterial strains found in natural environments. These bacteria consume asphaltene, aromatic components or benzene compounds, saturated hydrocarbons, and nitrogen, sulfur, and oxygen (NSO) compounds, converting them into harmless CO_2 and water. The oil zapper has a tendency to function at temperatures between 8°C and 40°C. Compared to other traditional treatments, using an oil zapper for biodegradation is 30%–40% more cost-effective. Implementing a modular system for the biodegradation process and testing for the treatment of hydrocarbon-contaminated sediments collected in Messina harbor (Italy), the study by Cappello et al., and Genovese et al., (Sharma et al., 2019).

10.2.8 Integration of Bioremediation Strategies

Synergistic approaches for addressing multiple contaminants simultaneously: Using plants to remediate a relatively new and developing method called phytoremediation employs plants to extract pollutants from contaminated soil and water. The rhizosphere's microbial community aids in the breakdown of organic contaminants in addition to promoting plant growth by encouraging the formation of roots (Elisaet al., 2017). The following are subprocesses of phytoremediation: Phytostabilization, which immobilizes pollutants to stop leaching and runoff; phytovolatilization, which releases pollutants into the atmosphere as volatile compounds after being absorbed from the soil; and rhizofiltration, which filters pollutants and excess nutrients through the root system. Phytoextraction: hazardous substances, chemicals, and heavy metals that are held in plant roots and prove to be poisonous to organisms in low quantities (Hector et al., 2011).

It is an inexpensive, less disruptive, environmentally benign, and aesthetically beautiful approach that may be used on a variety of pollutants. It is limited to areas where the contamination extends down to the plant roots; it is not appropriate for areas with high contaminant concentrations; it is slower than traditional methods; it works best during certain seasons; the plant biomass is disposed of after treatment; and root secretion may improve pollutant solubility and increase metal distribution into the soil environment (Narendrula et al., 2018).

10.2.9 Nano-Bioremediation

An innovative technique for getting rid of heavy metals and organic pollutants is called nano-bioremediation. This method uses bacteria, algae, fungi, and plants to create nanoparticles that are produced under controlled conditions and used for remediation. Remediation techniques can include bio, phyto, or nano-bioremediation, depending on the contaminants and live organisms employed.

10.2.10 Phycoremediation

Phycoremediation is a technique by which pollutants are absorbed and broken down by oxidation by algal species like *Chlorella* and certain fresh algae like *Chlorella vulgaris*. The extensive dispersion and high adaptability of both macro and microalgae make them suitable for use in the treatment of water and wastewater.

10.2.11 REMEDIATION USING BIOCHAR

Biochar is produced from agricultural wastes such as rice straw, sawdust, cassava stem, groundnut hull, cotton bract, and so on. It is also occasionally made from trash from homes. The waste material undergoes pyrolysis, a limited oxygen burn process. As a result, waste material is effectively used as a product in treatment technology. Biochar can be used in soil remediation to improve soil fertility, sequester carbon, and immobilize pollutants. It aids in the repair of problematic soil and raises the amount of organic contaminants in the soil. Biochar aids in the absorption of pollutants from the soil because of its high porosity and surface area (Ramaswamy et al., 2023).

10.2.12 CONSIDERATIONS FOR DESIGNING INTEGRATED BIOREMEDIATION STRATEGIES

There are financial benefits and certain drawbacks to using microbial techniques for bioremediation. Comprehending the economic aspects and constraints linked to these methodologies is imperative for their efficacious execution and extensive acceptance. Here are some important things to think about:

10.2.13 ECONOMICS

10.2.13.1 Cost-Effectiveness

Microbial treatments for potential cadmium contamination can be less expensive than typical remediation techniques that entail excavation, transportation, and disposal of Bioremediation: A Study of 129 Polluted Sources. Once established, bioremediation methods like phytoremediation and bioleaching might have lower operating and maintenance costs (Ahmad et al., 2017, Ali et al., 2021a, b).

10.2.13.2 Eco-Friendly and Sustainable

Compared to chemical treatments, microbial techniques are frequently seen to be more ecologically friendly and sustainable. They might therefore be more in line with the expanding need for environmentally friendly remedial techniques.

10.2.13.3 Possibility of Resource Recovery

A few microbial techniques, such as phytoextraction and bioleaching, have the potential to recover metals, which might be economically advantageous since they could be recycled or sold.

10.2.13.4 Long-Term Benefits

By allowing natural processes to continue cleaning the environment even after initial setup, microbial solutions can have long-term benefits by lowering the requirement for continuous interventions.

10.2.14 CONSTRAINTS

10.2.14.1 Site-Specific Factors

A number of site-specific factors, including soil type, climate, and the availability of beneficial microbes, affect the effectiveness of microbial methods. Certain tactics might not work well or be feasible in every situation.

10.2.14.2 Slow Cleanup Rates

When compared to chemical approaches, microbial strategies—especially plant-based ones like phytoremediation—can be slower. It can take longer for them to attain appreciable drops in Cd levels.

10.2.14.3 Inadequate Microbes

In order for bioremediation to be effective, certain habitats may not contain the metal-accumulating or metal-resistant microbes that are required. In certain situations, bioaugmentation may be required, which entails additional expense and complexity.

10.2.14.4 Uncertainty and Risk

Depending on the microbial community's complexity, the surrounding environment, and other variables, microbial techniques may be less dependable and produce inconsistent results. Regarding their performance and success, there can be doubts.

10.2.14.5 Requirements for Land and Space

The implementation of certain microbial techniques, like phytoremediation, may necessitate a significant amount of land or space, which may restrict their application in crowded or urban regions.

10.2.14.6 Regulatory Approvals

Adopting microbial techniques for bioremediation may include obtaining permissions and regulatory approvals, which could extend the time and expense of putting these technologies into practice.

10.2.14.7 Public Perception

Stakeholders and the general public who are not familiar with these novel approaches may oppose or be skeptical of the acceptance and adoption of microbial strategies. Continued research, development, and case studies proving the efficacy and financial feasibility of microbial techniques for Cd bioremediation are crucial to overcoming these obstacles and promoting their broad adoption.

Regulators, businesses, communities, and researchers can collaborate to ensure the successful and sustainable integration of microbial techniques into environmental remediation processes (Ahmad et al., 2017).

10.3　BIOREMEDIATION OF HEAVY METALS

10.3.1　Sources and Environment Impacts of Heavy Metal Contamination

In the environment, heavy metals are everywhere due to natural and anthropogenic activities. On the other hand, a few heavy metals regulate specific physiological processes in the body. Essential heavy metals that are found in nature enter the body through food, drink, and air and control a wide range of biological processes.

The most common heavy metals found in the environment are lead, nickel, copper, chromium, cadmium, arsenic, and mercury (Mitra et al., 2022). The presence of heavy metals even at low concentrations is still an environmental issue because of toxicity. Heavy metals alter soil fertility and quality, contaminate groundwater, cause biomagnification, and eventually destroy soil biota. Heavy metals also affect the plants in many ways. Small variations in the amount present above the allowable limit, whether brought on by man-made or natural causes, are extremely concerning since they can lead to major environmental issues and ensuing health issues. Agriculture-related practices such as using pesticides and herbicides, tainted irrigation water, and fertilizing with municipal waste are examples of anthropogenic sources of heavy metals. In addition, smoking, traffic emissions, farming waste disposal, mining activities, sewage discharge, and building materials like paints are examples of anthropogenic sources. Direct discharge of solid waste also contaminates water with heavy metals; direct inhalation of dusty soil and additives used in gasoline and lubricating oil contaminate the environment with heavy metals. Natural sources of heavy metal include different ways, as shown in Table 10.1 and Figure 10.3.

TABLE 10.1
Natural Sources of Heavy Metal Pollution (Nyiramigisha et al., 2022)

Sr. No.	Heavy Metal	Sources of Heavy Metal
1.	Cadmium (Cd)	1. Black shale has naturally occurred cadmium. 2. Parent material, marine sedimentary rocks, phosphates, and volcanic activity are the primary natural sources of Cd in the soil and atmosphere.
2.	Lead (Pb)	1. The sedimentary rocks, argillaceous rocks, and acidic igneous rocks. 2. Dolomites and limestone may also have an impact on the amount of lead in the soil. In addition, shale—particularly black shale—is another source of lead in the soil.
3.	Zinc (Zn)	1. Acidic granitic rocks and sedimentary rocks. 2. Clayey sediments and black shale Dolomite, limestone, and sandstone.
4.	Copper (Cu)	1. Copper exists naturally in various parent rocks, with basalts being an abundant source of the element. 2. Black shale and shale clay both naturally contain large amounts of Cu.
5.	Mercury (Hg)	1. The Earth's crust naturally degasses. 2. The argillaceous sediments, sedimentary rocks, and igneous rocks.

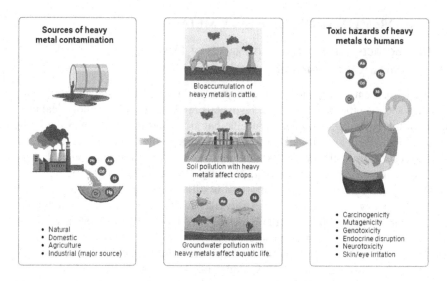

FIGURE 10.3　Different sources of heavy metal pollution.

10.3.2　Mechanism of Microbial-Mediated Heavy Metal Remediation

Remediation of heavy metals by microbes is vast but unexploited. A number of microorganisms utilize the pollutants as a source of energy for their metabolic functions. In the environment, fungus and bacteria aid in the breakdown or detoxification of toxic chemicals. The toxic (heavy) metals can be attached to extracellular polymeric substances on the biomass cell wall through a proton exchange mechanism with metal (Comte et al., 2008). The presence of different functional groups, such as carboxyl, amino, phosphoryl, and sulfa groups, which can be exchanged for ions, gives biomass surfaces a negative charge. Various mechanisms are used to perform bioremediation, including redox processes, adsorption, electrostatic attraction, precipitation, complexation, and ion exchange.

Metal mobilization can be initiated by microorganisms through redox reactions, and this can have an impact on bioremediation processes. Oxidation and reduction cycles are carried out on heavy metals such as Fe, As, Cr, and Hg. Transforming an element from its insoluble and stationary state in sediments into its soluble and mobile phase facilitates bioremediation. Detrimental effects on mobilization may also arise from the liberation of dangerous metal ions from their solid phase in sediments and their redistribution into the solution phase (Fomina et al., 2014). Heavy metals can therefore enter microbial metabolic pathways and are more bioavailable. The bacteria convert Hg (II) to Hg (0), which is more volatile and elemental (Wiatrowski et al., 2006). Additionally, microbial reduction can facilitate leaching from soil and increase the solubility of ions such as Fe (III) and As (V) by reducing them to Fe (II) and As (III), respectively (Bachate et al., 2012). Studies have shown transformation by microorganisms from several natural aquifers (Chang et al., 2010).

Aspergillus niger was used by Pokhrel and Vira Raghavan to eliminate As (V) and As (III) (Pokhrel et al., 2006). The process of biomethylation of toxic metals, which can alter the toxicity, volatility, and mobility of heavy metals, is significant in soil and water. Because volatile methylated species can be eliminated from cells, they also play a significant role in detoxification processes (Bolan et al., 2014). Because of their volatility and evaporation, soil loses dimethylmercury and alkyl arsines, which are the methylated derivatives of mercury and arsenic, respectively. The methyl group which is released into the soil is part of organic materials. The microbial breakdown of organic materials is another indirect process of metal mobilization that speeds up the release of these ions. It has been discovered that *Schizophyllum* commune releases heavy metals along with dissolved organic matter (Wengel et al., 2006). One significant method of chelating metal ions is the excretion of metabolites by microorganisms, such as amino acids and carboxylic acids.

10.3.3 STRATEGIES FOR ENHANCING MICROBIAL ACTIVITY IN HEAVY METAL-CONTAMINATED ENVIRONMENTS

The use of biotechnology in recent studies facilitates the growth of genetically modified organisms (GMOs), which is crucial for assessing them in comparison to strains that are wild-type. Because they have the right protein machinery, GMOs can use gene-controlling elements like promoters, binders, and terminators to absorb and regulate harmful substances. These organisms produce a protein that binds strongly to heavy metals, protecting against toxicity (Figure 10.4). The Mer operon gene, which affects both positive and negative transcription levels, regulates the removal of heavy metals. For the large-scale field-based treatment of contaminants, *P. fluorescens* strain HK44 was used.

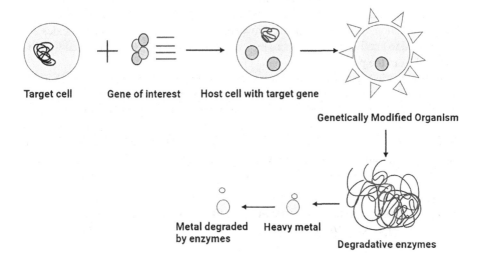

FIGURE 10.4 Role of GMOs in heavy metal degradation.

10.4   BIOREMEDIATION OF PESTICIDES

10.4.1   TYPES AND SOURCES OF PESTICIDE CONTAMINATION IN THE ENVIRONMENT

Based on nature and toxicity level, pesticides are of different types. The most commonly used approach to pesticide classification is their chemical structure, which is categorized into organochlorine, organophosphorus, carbamates, pyrethrin, and pyrethroids. Compared to other pesticides, pyrethrin and pyrethroids are less hazardous to mammals and less enduring in the environment. The table's chemical structure guides the classification (Rajmohan et al., 2020). Pesticides are categorized according to their intended use, roles, and the pest species they target. Insecticides, herbicides, rodenticides, bactericides, and fungicides are the primary classes. Remediation of pesticidal soil is prioritized because of the extensive use of pesticides by farms, which has an impact on the quality of groundwater (Mascarelli et al., 2013). The use of pesticides has been a key driver in boosting crop productivity.. Cotton leads all crops in terms of pesticide consumption "(45%), followed by rice (22%), vegetables (9%), plantation crops (7%), wheat (4%), and other crops (9%)". 6000 g/ha of pesticides are used in industrialized countries, but 600g/ha are utilized in poor nations like India. Through crop production, an estimated 4.6 million tons of pesticides are released into the ecosystem (Pradhan et al., 2022).

10.4.2   MICROBIAL DEGRADATION PATHWAYS FOR PESTICIDES

Using traditional chemical procedures for remediation and excavation of contaminated sludge carries a significant danger, and using microbes to break down pesticides is a difficult task. These days, biodegradable pesticides like organophosphate are preferred over resistant ones like organochlorines. Based on the site of bioremediation—ex situ versus in situ—biodegradation is further categorized. The aerobic method known as "in situ remediation" is done right at the contaminated site. A few in situ remediation techniques that can be used to get rid of pesticides are bioaugmentation, biosparging, bioaugmentation, biostimulation, and bioventing. In ex situ remediation, pesticides are transported to another location where they biodegrade while the contaminated water is removed from the polluted area.

In the process, pesticides are removed from the environment by bacteria using them as co-substrates in their metabolic activity. On the other hand, because they function as strong catalysts and can alter the chemical makeup and toxicity of pollutants to less hazardous forms, bacteria and enzymes are more efficient in the breakdown of organic pollutants (Zawierucha et al., 2011). The microorganisms listed in Table 10.2 that are capable of degrading pesticides are bacteria and fungi (Gangola et al., 2018; Khajezadeh et al., 2020; Luo et al., 2018; Purnomo et al., 2020).

10.4.3   BIOREMEDIATION APPROACHES FOR PESTICIDES-CONTAMINATED SOILS AND WATER BODIES

To optimize the efficiency of removal in the bioremediation process, pH, water content, temperature, oxygen, and nutrients must be regulated (Raffa et al., 2021).

TABLE 10.2
List of Microbes Which Are Capable for Degrading Pesticides

Sr. No.	Pesticides	Microorganisms
1.	Deltamethrin	*Streptomyces rimosus*
2.	Fentopropathrin	*Rhodopseudomonas palustris*
3.	Tebuconazole	*Serratia marcescens*
4.	DDT	*Fomitopsis pinicola Ralstonia pickettii*
5.	Phorate	*Brevibacterium frigoritolerans Bacillus aerophilus Pseudomonas fulva*
6.	Acetochlor	*Tolypocladium geodes Cordyceps*

- **Natural Attenuation:** It is a process where the pollutant is reduced by indigenous microorganisms present in the soil. The natural processes include biological degradation, volatilization, dispersion, dilution, radioactive decay, and sorption of the contaminant onto the organic matter and clay minerals in the soil.
- **Biostimulation:** In this process, nutrients such as phosphorous, nitrogen, and carbon are added to promote the growth of indigenous microorganisms. Nutrient addition stimulates the biodegradation process, increasing the diversity of microbial species. Betancur-Corredor et al. have analyzed the degradation of Dichlorodiphenyltrichloroethane (DDT), Dichlorodiphenyldichloroethane (DDD), and Dichlorodiphenyldichloroethylene (DDE), stimulating the microbial population and adding a surfactant. Throughout the process, the amount of nutrients delivered was controlled to prevent any reduction in microbial activity.
- **Bioaugmentation:** By increasing the microbial diversity, this procedure entails introducing microbial consortia or a single strain into the soil. Doolotkeldieva et al. discovered that the soils under study had a variety of bacterial strains. Next, they examined the insecticide aldrin's breakdown. Aldrin is a dispersed chlorinated hydrocarbon. The findings have shown that bacteria strains, specifically *Bacillus polymyxa* and *Pseudomonas fluorescens*, possessing particular genes (cytochrome P450) are able to break down aldrin in a comparatively short amount of time. The bioaugmentation treatment was assessed by Odukkathil et al. in an experimental test setup in a glass column with a 4500 cm^3 capacity. In bottom soil, pesticide concentration is high compared to central soil due to microbial activity, which favors the degradation of pollutants (Odukkathil et al., 2016).
- **Composting:** The process involves incorporating non-hazardous organic amendment into the contaminated soil to encourage the growth of a bacterial and/or fungal population that can break down the pesticides via a co-metabolic pathway. This method demonstrated low pesticide concentrations.
- **Slurry Bioreactor:** The contaminated soil is combined with wastewater in a slurry bioreactor to create a slurry that has aqueous suspensions ranging from 10% to 30% W/V. A bioreactor can function both aerobically and anaerobically. The anaerobic biodegradation of organochlorine insecticides

was investigated by Baczynski et al. Anaerobic treatment of soil contaminated with γ-hexachlorocyclohexane, methoxychlor, o,p'-, and p,p'-DDT was carried out using methanogenic granular sludge as the inoculum. At all measured temperatures—12°C, 22°C, and 30°C—the contaminants were substantially eliminated without experiencing a lag period. Pesticide removal rates are decreased at low temperatures.

10.4.4   CHALLENGES AND CONSIDERATIONS SPECIFIC TO PESTICIDES BIOREMEDIATION

When people are exposed to pesticides at long term, sublethal levels, they can cause a variety of chronic ailments that impact both the environment and humans in underdeveloped countries. Sadly, underdeveloped countries lack the resources to conduct worker and application exposure studies that could prevent or significantly reduce poisonings and fatalities. Understanding the factors that lead to poisoning is necessary for the prevention of pesticide poisoning. To reduce the number of acute and long-term damage brought on by pesticides, governments, environmental protection organizations, farmers, and pesticide manufacturers must work together. Health professionals and manufacturers bear the responsibility of implementing appropriate measures to ensure the effective management of pesticides by means of stringent legislation and toxicity limits.

This entails spreading knowledge about the responsible use of pesticides only, when necessary, farmers donning the requisite safety masks, and using the right applicators to reduce the impacts of potentially harmful substances. Furthermore, integrated pest management strategies incorporating biological and cultural interventions must be implemented in all developing countries. With proper preparation and commitment to reducing the usage of toxic pesticides, this is achievable. In certain regions of India, composting (both vermicomposting and aerobic composting) and waste-to-energy (WTE) (incineration, palletization, and biomethanation) are the two most cutting-edge methods of disposing of trash. WTE is a relatively recent idea that hasn't been widely implemented in urban solid waste management projects.

10.4.5   CASE STUDIES ILLUSTRATING SUCCESSFUL REMEDIATION OF PESTICIDES-CONTAMINATED SITES USING BIOREMEDIATION TECHNIQUES

10.4.5.1   Bioremediation Technique: Land Farming

"Pesticides: Hexachlorocyclohexane (HCH) isomers (insecticides)

Description: Contaminated soil with HCH isomers (>5000 mg/kg) derived from lindane production was studied in the field for 11 months, setting up two plots (each 2 m × 10 m). The α- and γ-HCH isomers were decreased by 89% and 82% of the initial concentration, respectively. The concentration of the most persistent β-isomer remained essentially unaffected" (Rubino et al., 2007).

10.4.5.2   Bioremediation Technique: Bioaugmentation

"Pesticides: Myclobutanil, tetraconazole, and flusilazole.

Description: Experimental tests were conducted on vineyard plots. In the crops, an agricultural formulation of pesticides by foliar spray was applied. After one h of pesticide application, vines were sprayed with a suspension of four *Bacillus* strains. DR-39, CS-126, TL-171, and TS-204 were tested. Residue analysis of field samples showed 87.4% and >99% degradation of myclobutanil and tetraconazole, respectively, by the strain DR-39, and 90.8% degradation of flusilazole by the strain CS-126 after 15–20 days of treatment" (Salunkhe et al., 2015).

10.4.5.3 Bioremediation Technique: Biostimulation

"Pesticides: Organochlorine pesticides: toxaphene; DDT; DDE; DDD; endosulfan II; γ-chlordane; α-chlordane; dieldrin.

Description: The Mantegani Property is a 0.8-acre area treated with soil amendment to help the indigenous bacteria to metabolize the pesticides. High concentrations of DDT and dieldrin were present. After treatment, DDT was degraded by 97% and dieldrin by 73%, while the concentrations of other Oral Contraceptive Pills (OCPs) were below their preliminary remediation goals" (Sacramento et al., 2010).

10.5 CHALLENGES AND FUTURE DIRECTIONS

10.5.1 Challenges

10.5.1.1 Microbial Diversity and Efficacy

- **Microbial Restrictions:** Not all polluted locations contain the necessary microbial populations for efficient bioremediation. In the event that certain hydrocarbon-degrading microorganisms are lacking, bioaugmentation—which increases complexity and expense—may be required.
- **Microbial Consistency:** Variations in the makeup of the microbial community and environmental factors might cause inconsistent microbial degradation efficacy, which can have unpredictable results.
- **Complexity of Pesticide Structures:** Especially for organochlorine and organophosphorus compounds, pesticides frequently have intricate chemical structures. Effective bioremediation may face substantial difficulties due to these materials' potential resistance to microbial breakdown. Certain pesticides, like DDT, have a significant potential for toxicity and persistence in the environment. Because of their lengthy half-lives, they are difficult to fully decay, which might cause long-term pollution problems.

10.5.1.2 Environmental Restrictions

- **Site-specific Factors:** The success of bioremediation is greatly influenced by the kind of soil, the climate, and the availability of nutrients. The applicability of a one-size-fits-all strategy is limited by the large variations in these characteristics across different sites.
- **Slow Cleanup Rates:** The fact that chemical procedures frequently work more quickly than microbial and plant-based techniques, including phytoremediation, might be a disadvantage for locations that need to be decontaminated quickly.

- **Partial Degradation:** A few bioremediation techniques could produce intermediate compounds that are just as toxic—if not more so—than the original pesticide. A crucial difficulty is ensuring full reduction to non-toxic end products.

10.5.1.3 Economic and Technical Barriers

- **Cost Considerations:** Although microbial approaches can be very economical in the long run, they can be very expensive in the setup phase, particularly for bioaugmentation and biostimulation. Operational complexity is further increased by maintaining an ideal nutritional balance to prevent microbial inhibition.
- **Regulatory and Public Perception:** It can take some time for novel microbiological procedures to receive regulatory approval. Implementation may also be hampered by public mistrust and hostility to GMOs, which are employed in bioaugmentation.

10.5.1.4 Operating Difficulties

- **Nutrient Imbalance:** Excessive nutrient delivery during biostimulation might result in imbalances that prevent microbiological activity; therefore, careful control is essential.
- **Space and Land Requirements:** The application of techniques such as phytoremediation is restricted in urban or densely inhabited areas due to their substantial land requirement.

10.5.2 FUTURE DIRECTION

10.5.2.1 Improving Microbiological Methods

- **Genetic Engineering:** New developments in this field have the potential to generate more resilient microbial strains that can break down a wider variety of hydrocarbons and thrive in a variety of environmental settings. The process of creating bacteria with improved pesticide-degrading abilities through genetic engineering. This may entail inserting genes into resilient microbial hosts that encode particular degradative enzymes.
- **Microbial Consortia:** It is possible to improve degrading efficiency and resilience to environmental stressors by creating artificial microbial consortia that resemble natural populations.

10.5.2.2 Combined Bioremediation Techniques

- **Synergistic Approaches:** Complex contaminations can be more successfully addressed by combining physical and chemical procedures with bioremediation techniques like bioaugmentation and biostimulation.
- **Nano-Bioremediation:** By utilizing nanotechnology to produce bio-nano-composites, contaminants can be absorbed and broken down more effectively, offering a fresh and effective remediation strategy. Pesticides can

be more bioavailable thanks to nanoparticles, which also make it easier for microorganisms to absorb and degrade them.

10.5.2.3 Monitoring and Optimization
* **Advanced Monitoring Tools:** By utilizing biosensors in real-time monitoring systems, NELs and microbial activity may be optimized, leading to more reliable and efficient bioremediation.
* **Modeling and Simulation:** By creating prediction models to mimic the steps involved in bioremediation, treatment methods can be optimized and field applications can be guided.

10.5.2.4 Economic and Environmental Aspects to Consider
* **Sustainable Practices:** Increasing the adoption of eco-friendly techniques will increase the overall sustainability of bioremediation. Examples of these techniques include applying biochar to improve soil and immobilize pollutants.
* **Resource Recovery:** Adopting bioremediation techniques might be further encouraged by investigating the financial potential of resource recovery from contaminated locations, such as metal recovery by bioleaching.

10.5.2.5 Public Engagement and Policy
* **Regulatory Support:** Broader adoption can be facilitated by streamlining regulatory procedures and offering financial incentives for bioremediation projects. Public education can help increase public acceptance and support for microbial approaches by educating people about the advantages and safety of bioremediation.

10.6 CONCLUSIONS

When it comes to the repair of environmental degradation brought on by pesticides, heavy metals, and petroleum hydrocarbons, bioremediation stands out as a viable and sustainable method. The review highlights a number of bioremediation approaches that have been successful in removing these toxins, demonstrating a diversity of methods suited to various pollutants and environmental circumstances. Even though bioremediation of pesticides, heavy metals, and petroleum hydrocarbons has advanced significantly, further research and technical developments are needed to get past present obstacles and improve the efficacy of these techniques. In order to create comprehensive solutions for environmental contamination, future strategies should concentrate on multidisciplinary techniques that incorporate biological, chemical, and physical methodologies. For bioremediation methods to be implemented successfully, it will also be essential to address public concerns and regulatory obstacles. We can increase environmental health and resilience by pursuing these tactics further and achieving more resilient and sustainable remediation solutions for contaminated environments.

REFERENCES

Abbasian, F., Lockington, R., Mallavarapu, M., Naidu, R., 2015. A comprehensive review of aliphatic hydrocarbon biodegradation by bacteria. *Applied Biochemistry and Biotechnology*, 176(3), 670–699. doi: 10.1007/s12010-015-1603-5.

Ahmad, I., Akhtar, M.J., Jadoon, I.B.K., Imran, M., Imran, M., Ali, S. 2017. Equilibrium modeling of cadmium biosorption from aqueous solution by compost. *Environmental Science and Pollution Research*, 24, 5277–5284.

Alaidaroos, B. 2023. Advancing eco-sustainable bioremediation for hydrocarbon contaminants: challenges and solutions. *Processes*, 11. 3036. doi: 10.3390/pr11103036.

Ali, J., Ali, F., Ahmad, I., Rafique, M., Munis, M.F.H., Hassan, S.W., et al., 2021a. Mech-anistic elucidation of germination potential and growth of Sesbania sesban seedlings with Bacillus anthracis PM21 under heavy metals stress: an in vitro study. *Ecotoxicology and Environmental Safety*, 208, 111769.

Ali, J., Wang, X., Rafique, M., Ahmad, I., Fiaz, S., Munis, M.F.H., Chaudhary, H.J. 2021b. Phytoremediation of cadmium contaminated soil using Sesbania sesban L. in association withbacillus anthracis PM21: a biochemical analysis. *Sustainability*, 13(24), 13529.

Bachate, S.P., Khapare, R.M., Kodam, K.M, 2012. Oxidation of arsenite by two β-proteobacteria isolated from soil. *Applied Microbiology and Biotechnology*, 93(5), 2135–2145.

Bala, S., Garg, D., Thirumalesh, BV., Sharma, M., Sridhar, K., Inbaraj, BS., Tripathi, M, 2022. Recent Strategies for Bioremediation of Emerging Pollutants: A Review for a Green and Sustainable Environment. *Toxics*. 2022 Aug 19; 10(8): 484.

Bolan, N., Kunhikrishnan, A., Thangarajan, R., Kumpiene, J., Park, J., Makino, T., Kirkhamm M.B., Scheckel, K. 2014. Remediation of heavy metal(loid)s contaminated soils – to mobilize or to immobilize? *Journal of Hazardous Materials*, 15(266), 141–166.

Chang, J.S., Yoon, I.H., Lee, J.H., Kim, K.R., An, J., Kim, K.W, 2010. Arsenic detoxification potential of aox genes in arsenite-oxidizing bacteria isolated from natural and constructed wetlands in the Republic of Korea. *Environmental Geochemistry and Health*, 32(2), 95–105. doi: 10.1007/s10653-009-9268-z.

Chettri, B., Singha, N.A., Singh, A.K. 2021. Efficiency and kinetics of Assam crude oil degradation by Pseudomonas aeruginosa and Bacillus sp. *Archives of Microbiology*, 203, 5793–5803. doi:10.1007/s00203-021-02567-1.

Chikere, C.B., Okpokwasili, G.C., Chikere, B.O., 2011. Monitoring of microbial hydrocarbon remediation in the soil. *3 Biotech*, 1(3), 117–138. doi: 10.1007/s13205-011-0014-8.

Comte, S., Guibaud, G., Baudu, M., 2008. Biosorption properties of extracellular polymeric substances (EPS) towards Cd, Cu and Pb for different pH values. *Journal of Hazardous Materials*, 151(1), 185–193.

Conrad, R., 2020. Importance of hydrogenotrophic, acetolactic and methylotrophic methanogenesis for methane production in terrestrial, aquatic and other anoxic environments: a mini review. *Pedosphere*, 30(1), 25–39. doi: 10.1016/S1002–0160(18)60052-9.

Costa, J.L., Silva, L.G., Kato, M.T., 2024. Use of nitrate, sulphate, and iron (III) as electron acceptors to improve the anaerobic degradation of linear alkylbenzene sulfonate: effects on removal potential and microbiota diversification. *Environmental Science and Pollution Research*, 2024. doi: 10.1007/s11356-024-33158-4.

Cui, J.Q., He, Q.S., Liu, M.H., Chen, H., Sun, M.B., Wen, J.P. 2020. Comparative study on different remediation strategies applied in petroleum-contaminated soils. *International Journal of Environmental Research and Public Health*, 17, 1606. doi:10.3390/ijerph17051606.

Denkler, L.M., Shekar, M.A., Ngan, T.S.J., Schnakenburg, G.A. 2024. General iron-catalyzed decarboxylative oxygenation of aliphatic carboxylic acids. *Angewandte Chemie International Edition*, 2024. doi: 10.1002/anie.202403292.

Dindar, E., Olcay, F., Sagban, T., Baskaya, H.S. 2013. Bioremediation of petroleum-contaminated soil petrol ile kirlenmiş toprakların biyoremediasyonu. *Journal of Environmental Biology*, 7, 39–47. doi:10.3390/ijms19113373.

Doolotkeldieva, T., Konurbaeva, M., Bobusheva, S. 2018. Microbial communities in pesticide-contaminated soils in Kyrgyzstan and bioremediation possibilities. *Environmental Science and Pollution Research*, 2018(25), 31848–31862.

Elisa, F., Anna, M., Flavio, M., Antonella, F., Giovanni Dal, C. 2017. The potential of genetic engineering of plants for the remediation of soils contaminated with heavy metals. *Plant, Cell & Environment*, 41(5), 1201–1232.

Fomina, M., Gadd, G.M. 2014. Biosorption: current perspectives on concept, definition and application. *Bioresource Technology*, 160, 3–14.

Gangola, S., Sharma, A., Bhatt, P., Khati, P., Chaudhary, P. 2018. Presence of esterase and laccase in Bacillus subtilis facilitates biodegradation and detoxification of cypermethrin. *Scientific Reports*, 8, 1–11. doi: 10.1038/s41598-018-31082-5.

Gaur, V.K., Gupta, S., Pandey, A. 2021. Evolution in mitigation approaches for petroleum oil-polluted environment: recent advances and future directions. *Environmental Science and Pollution Research*, 29. doi: 10.1007/s11356-021-16047-y.

Gogoi, G., Nath, J.K., Hoque, N., Bania, K.K., 2022. Single and multiple site Cu(II) catalysts for benzyl alcohol and catechol oxidation reactions. *Applied Catalysis*, 2022. doi: 10.1016/j.apcata.2022.118816.

Guarino, C., Spada, V., Sciarrillo, R. 2017. Assessment of three approaches of bioremediation (Natural Attenuation, Landfarming and Bioagumentation – assistited Landfarming) for a petroleum hydrocarbon contaminated soil. *Chemosphere*, 170, 10–16. doi: 10.1016/j.chemosphere.2016.11.165.

Haohao Sun, Takashi Narihiro, Xueyan Ma, Xu-Xiang Zhang, Hongqiang Ren, Lin Ye, 2019. Diverse aromatic-degrading bacteria present in a highly enriched autotrophic nitrifying sludge, *Science of the Total Environment*, 666, 245–251. ISSN 0048–9697. doi: 10.1016/j.scitotenv.2019.02.172.

Haritash, A.K., Kaushik, C.P., 2009. Biodegradation aspects of polycyclic aromatic hydrocarbons (PAHs): a review. *Journal of Hazardous Materials*, 169(1), 1–15. doi: 10.1016/j.jhazmat.2009.03.137.

Hector, M.C., Michael, W.H., Evangelou, B.H., Robinson, Rainer, S. 2011. Environmental impacts of pesticides and fertilizers. *The Scientific World Journal*, Article ID-173829.

Higashioka, Y., H. Kojima, M. Fukui. 2012. Isolation and char-acterization of novel sulfate-reducing bacterium capable of anaerobicdegradation of p-xylene. *Microbes and Environment*, 27, 273–277.

Kamani, H., Baniasadi, M., Abdipour, H., Mohammadi, L., Rayegannakhost, S., Moein, H., Azari A. 2023. Health risk assessment of BTEX compounds (benzene, toluene, ethylbenzene and xylene) in different indoor air using Monte Carlo simulation in zahedan city, Iran. *Heliyon*, 9, e20294.

Khajezadeh, M., Abbaszadeh-Goudarzi, K., Pourghadamyari, H., Kafilzadeh, F. 2020. A newly isolated Streptomyces rimosus strain capable of degrading deltamethrin as a pesticide in agricultural soil. *Journal of Basic Microbiology*, 60, 435–443. doi: 10.1002/jobm.201900263.

Kolukirik M., Ince O., 2011. Nutrient enhanced bioremediation of petroleum hydrocarbons in anoxic marine sediments.*A/Z ITU journal of Faculty of Architecture*, 21(1), 55–65.

Kvenvolden, K.A., Cooper, C.K. 2003. Natural seepage of crude oil into the marine environment. *Geo-Marine Letters*, 23, 140–146. doi:10.1007/s00367-003-0135-0.

Laban, N.A., Dao, A., Semple, K., Foght, J., 2014. Biodegradation of C7 and C8iso-alkanes under methanogenic conditions. *Environmental Microbiology*, 16(11), 3317–3329.

Lawrence, S.J., 2006. Description, properties, and degradation of selected volatile organic compounds detected in ground water – a review of selected literature no. 2006-1338. *U.S. Geological Survey*.

Li, Y., Liu, Y., Guo, D., Dong, H., 2024. Differential degradation of petroleum hydrocarbons by Shewanellaputrefaciens under aerobic and anaerobic conditions. *Frontiers in Microbiology*, 15. doi: 10.3389/fmicb.2024.1389954.

Luo, X., Zhang, D., Zhou, X., Du, J., Zhang, S., Liu, Y. 2018. Cloning and characterization of a pyrethroid pesticide decomposing esterase gene, Est3385, from Rhodopseudomonas palustris PSB-S. *Scientific Reports*, 8, 1–8. doi: 10.1038/s41598-018-19373-3.

Martelletti, S. 2015. Constraints in lowland forest restoration: the acorn predation issue, in Sostenere il pianeta, boschi per la vita-Ricerca e innovazione per la tutela e la valorizzazione delle risorse forestali.

Martínez Álvarez, L.M., Ruberto, L.A.M., Lo Balbo, A., Mac Cormack, W.P. 2017. Bioremediation of hydrocarbon-contaminated soils in cold regions:development of a pre-optimized biostimulation biopile-scale field assay inAntarctica. *Science of the Total Environment*, 590–591, 194–203. doi: 10.1016/j.scitotenv.2017.02.204.

Mawardi, M., Indrawati, A., Wibawan, W.T., Lusiastuti, A., 2023. Antimicrobial susceptibility test and antimicrobial resistance gene detection of extracellular enzyme bacteria isolated from tilapia (Oreochromisniloticus) for probiotic candidates. *Veterinary World*, 16(2), 264–271. doi: 10.14202/vetworld.2023.264-271.

Mekonnen, B.A., Aragaw, T.A., Genet, M.. 2024. Bioremediation of petroleum hydrocarbon contaminated soil: a review on principles, degradation mechanisms, and advancements. *Frontiers in Environmental Science*, 12. doi: 10.3389/fenvs.2024.1354422.

Mitra, S, Chakraborty, A.J., Tareq, A.M., Emran, T.B., Nainu, F., Khusro, A., Idris, A.M., Khandaker, M.U., Osman, H., Alhumaydh, F.A., Simal-Gandara, J, 2022. Impact of heavy metals on the environment and human health: novel therapeutic insights to counter the toxicity. *Journal of King Saud University-Science*, 34(3), 101865.

Mrozik, A., Piotrowska-Seget, Z. 2010. Bioaugmentation as a strategy forcleaning up of soils contaminated with aromatic compounds. *Microbiological Research*, 165, 363–375. doi: 10.1016/j.micres.2009.08.001.

Narendrula Kotha, R., Mehes Smith, M., Nkongolo, K.K. 2018. Microbial response to soil contamination: An environmental approach. *International Journal of Environmental Bioremediation & Biodegradation*, 6(1), 1–7.

Nazifa, T., Ahmad, M.A., Hadibarata, T., Salmiati, S., Aris, A.. 2018. Bioremediation of diesel oil spill by filamentous fungus Trichoderma reesei H002 in aquatic environment. *International Journal of Integrated Engineering*, 10, 14–19. doi: 10.30880/ijie.article/view/2369/1952.

Nyiramigisha, P., Komariah, S. 2021. Harmful impacts of heavy metal contamination in the soil and crops grown around dumpsites. *Reviews in Agricultural Sciences*, 9, 271–282.

Odukkathil, G.; Vasudevan, N. 2016. Residues of endosulfan in surface and subsurface agricultural soil and its bioremediation. *Journal of Environmental Management*, 165, 72–80.

Omokhagbor Adams, G., Tawari Fufeyin, P., Eruke Okoro, S., Ehinomen, I. 2020. Bioremediation, biostimulation and bioaugmention: a review. *International Journal of Environmental Bioremediation & Biodegradation*, 3, 28–39.

Ossai, I.C. 2020. Remediation of soil and water contaminated with petroleum hydrocarbon: a review. *Environmental Technology & Innovation*, 17, 100526.

Pandit, M.A., 2022. *Anaerobes and anaerobic processes*. New Delhi: New India Publishing Agency.

Pokhrel, D., Viraraghavan, T. 2006. Arsenic removal from an aqueous solution by a modified fungal biomass. *Water Research*, 40(3), 549–552.

Purnomo, A.S., Sariwati, A., Kamei, I. 2020. Synergistic interaction of a consortium of the brown-rot fungus Fomitopsis pinicola and the bacterium Ralstonia pickettii for DDT biodegradation. *Heliyon*, 6, e04027. doi: 10.1016/j.heliyon.2020.e04027.

Raffa, C.M., Chiampo, F. 2021. Bioremediation of agricultural soils polluted with pesticides: a review. *Bioengineering*, 8, 92. doi: 10.3390/bioengineering8070092.

Rajmohan, K.S., Chandrasekaran, R., Varjani, S. 2020. A review on occurrence of pesticides in environment and current technologies for their remediation and management. *Indian Journal of Microbiology*, 60(2), 125–138.

Ramaswamy, G., Balu, S., Kanmani, S. 2023. Emerging contaminants in the environment and bioremediation control strategies – a review. *IOP Conference Series: Earth and Environmental Science*, 1258, 012002. doi:10.1088/1755-1315/1258/1/012002.

Sharma, N.. 2019. Microbes and their secondary metabolites: agents in bioremediation of hydrocarbon contaminated site. *Archives of Petroleum & Environmental Biotechnology*, 4: 151.

Shi, L., Liu, Z., Yang, L., Fan, W. 2022. Effect of plant waste addition as exogenous nutrients on microbial remediation of petroleum – contaminated soil. *Annual Review of Microbiology*, 72, 22. doi: 10.1186/s13213-022-01679-3.

Singha, L.P., Pandey, P. 2021. Rhizosphere assisted bioengineering approaches for the mitigation of petroleum hydrocarbons contamination in soil. *Critical Reviews in Biotechnology*, 41, 749–766. doi: 10.1080/07388551.2021.1888066.

Subhadarsini Pradhan, S., Basana Gowda, G., Adak, T., Guru-Pirasanna-Pandi, G.B., Patil, N., Annamalai, M., et al., 2022. Pesticides occurrence in water sources and decontamination techniques. In *Pesticides – Updates on Toxicity, Efficacy and Risk Assessment*. IntechOpen. doi: 10.5772/intechopen.103812.

Trejos Delgado, C., Gloria, E., Cadavid, R., Angelina Hormaza, A., Edison, A.A., Leonardo Barrios, Z. 2020. Oil bioremediation in a tropical contaminated soil using a reactor. *Anais da Academia Brasileira de Ciências*, 92(2), 20181396.

Usman, N., Atta, H., Tijjani, M.B. 2020. Biodegradation studies of benzene, toluene, ethylbenzene and xylene (BTEX) compounds by Gliocladium sp. and Aspergillus terreus. *Journal of Applied Sciences and Environmental Management*, 24, 1063–1069. doi: 10.4314/jasem.v24i6.19.

Weelink, S.A.B., M.H.A. van Eekert, A.J.M. Stams. 2010. Degradation of BTEX by anaerobic bacteria: physiology and application. *Reviews Environmental Science and Biotechnology*, 9, 359–385.

Wengel, M., Kothe, E., Schmidt, C.M., Heide, K., Gleixner, G. 2006. Degradation of organic matter from black shales and charcoal by the wood-rotting fungus Schizophyllum commune and release of DOC and heavy metals in the aqueous phase. *Science of the Total Environment*, 367(1), 383–393.

Wiatrowski, H.A., Ward, P.M., Barkay, T., 2006. Novel reduction of mercury (II) by mercury-sensitive dissimilatory metal reducing bacteria. *Environmental Science & Technology*, 40(21), 6690–6696.

Xue, J., Yu, Y., Bai, Y., Wang, L. and Wu, Y., 2015. Marine oil-degrading microorganisms and biodegradation process of petroleum hydrocarbon in marine environments: a review. *Current Microbiology*, 71(2), 220–228. doi: 10.1007/s00284-015-0825-7.

Yadav, A. N., Rastegari, A. A., Yadav, N. 2020. *Microbiomes of extreme environments*. Boca Raton, FL: CRC Press, Taylor & Francis Group.

Yoshikawa, M., Zhang, M., Toyota, K. 2017. Biodegradation of volatile organic compounds and their effects on biodegradability under co-existing conditions. *Microbes and Environments*, 32. doi: 10.1264/jsme2.ME16188.

Zafra, G., Cortés-Espinosa, D.V., 2015. Biodegradation of polycyclic aromatic hydrocarbons by Trichoderma species: a mini review. *Environmental Science and Pollution Research*, 22(24), 19426–19433. doi: 10.1007/s11356-015-5602-4.

Zeaiter, M., Belot, L., Valerie, C., Schlatter, U., 2024. Acetyl-CoA synthetase (ACSS2) does not generate butyryl- and crotonyl-CoA. *Molecular Metabolism*, 2024. doi: 10.1016/j.molmet.2024.101903.

Zhang, Y., Zhao, Z., Li, Y., 2020. Direct interspecies electron transfer in anaerobic digestion: research and technological application. *Chinese Science Bulletin* 65(26), 2820–2834. doi: 10.1360/TB-2020-0661.

Zhang, C., Meckenstock, R., Weng, S., Wei, G., Hubert, C., Wang, J. -H., Dong, X.. 2020. Marine sediments harbor diverse archaea and bacteria with the potential for anaerobic hydrocarbon degradation via fumarate addition. doi: 10.21203/rs.3.rs-71489/v1.

Zindel, U., Freudenberg, W., Rieth, M., Widddel, F., 1988. Eubacteriumacidaminophilum sp. nov., a versatile amino acid-degrading anaerobe producing or utilizing H_2 or formate. *Archives of Microbiology* 150(3), 1988, 254–266. doi: 10.1007/BF00407789.

Index